Why Nations Cooperate

Why Nations Cooperate

Circumstance and Choice in International Relations

ARTHUR A. STEIN

Cornell University Press

ITHACA AND LONDON

First published 1990 by Cornell University Press.

International Standard Book Number 0-8014-2417-8 (cloth)
International Standard Book Number 0-8014-9781-7 (paper)
Library of Congress Catalog Card Number 90-55132

Printed in the United States of America

Librarians: Library of Congress cataloging information
appears on the last page of the book.

⊛ The paper used in this book meets the minimum requirements
   of the American National Standard for Information Sciences—
   Permanence of Paper for Printed Library Materials, ANSI Z39.48-1984.

For Amy

Contents

Preface

Peace, it seems, is breaking out all over. We appear to be entering an era of hope, of the promise of international cooperation. But exuberence waxes and wanes, and such periods have alternated with times when international tensions have engendered omnipresent fears of likely war between the major powers. Not so long ago, after all, Americans worried about the morality of refusing to share their fallout shelters. More recently, a U.S. president depicted the Soviet Union as evil incarnate, and our grade schools developed a "nuclear curriculum."

Like the popular mood, although not always in sync with it, the academic outlook also has cycles. In some periods, scholars stress the cooperative components of international relations. In the 1920s, consistent with the euphoria of that time, they pointed to the potential of international law for bringing world order. But as the Cold War began in the late 1940s, in the wake of fascism and World War II, analysts emphasized the inherently conflictual nature of global politics.

Ironically, the early 1980s witnessed a divergence in popular and scholarly sentiment. In the United States and Western Europe a bleak public mood greeted the modernization of U.S. nuclear weapons delivery systems and vast increases in U.S. defense spending. And yet even as arms control negotiations broke down and fear of nuclear war grew, some scholars suggested not only that international cooperation was possible, but that it should evolve without negotiations and without international agreements or international institutions. In scholarly parlance, cooperation was achievable and sustainable even under anarchy.

Nonetheless, few if any foresaw that the collapse of American-Soviet détente would soon be followed by the end of the Cold War, the collapse of the Warsaw Pact, and the crumbling of the Berlin Wall.

This book contrasts, formally assesses, and illustrates realist and liberal contentions about international relations. The former stress the conflictual nature of international politics; the latter emphasize cooperation. Yet, I argue, both derive their contradictory conclusions from a shared set of core assumptions. The various chapters of this work take particular strands of realist and liberal thought and assess their implications for international cooperation and conflict. I demonstrate that many of the typically proffered assumptions of realism are inadequate to sustain an expectation of conflict. Other parts of the realist argument do hold up, but only under certain circumstances. And although many liberal views of the possibilities for cooperation in an anarchic world are also sustained, several liberal conclusions are significantly modified. It is time to transcend the liberal/realist debate and focus on the circumstances that underlie international interactions.

Two of the chapters are revised and expanded versions of previously published articles. Chapter 2 appeared originally as "Coordination and Collaboration: Regimes in an Anarchic World" in *International Organization* 36 (Spring 1982) and is reprinted with the permission of MIT Press. Chapter 3 first appeared as "When Misperception Matters" in *World Politics* 34 (July 1982), copyright © 1982 by Princeton University Press, reprinted with permission of Princeton University Press.

Over the years a number of people have commented on all or parts of the various incarnations of this manuscript. My thanks to Amy Davis, John Ferejohn, Roger Haydon, Robert Jervis, Robert Keohane, Edward Kolodziej, Stephen Krasner, Deborah Larson, Robert Pahre, Paul Papayoanou, Richard Rosecrance, Bruce Russett, Lars Skalnes, Cherie Steele, Thomas Willett, and the graduate students in Richard Rosecrance's advanced international relations theory course.

Special thanks are due the regulars in the Tuesday Political Economy lunch group. The lunch has been an excellent place for trying out ideas and working through readings of interest. My appreciation for the intellectual sustenance goes to all who have participated over the years.

I have been fortunate to have been surrounded at UCLA by excellent colleagues, always prepared to schmooze. Parts of the book began as a dialogue with my former colleagues Robert Jervis and Stephen Krasner. The hallway conversation has continued with James DeNardo, Jeffry Frieden, David Lake, Deborah Larson, Ronald Rogowski, Richard Rosecrance, George Tsebelis, and Michael Wallerstein. I offer my thanks to Marvin Hoffenberg for his unflagging good cheer and encouragement. Also to Michael Intriligator for sustaining the Jacob Marschak Interdisciplinary Colloquium on Mathematics in the Behavioral Sciences, which brings together people working on related issues who, all too often, are unaware of one another's efforts. Deserving of special acknowledgment are two people whose scholarship and humanity I admire, Jack Hirshleifer and Harold Kelley.

Along the way, I had a lot of help. To John Steinbruner and those in the Foreign Policy Studies program at The Brookings Institution go my thanks for providing me a congenial place to work away from home. Words of appreciation are due also to Peter Davis, Cherie Steele, and especially Elizabeth Bailey for research assistance. My thanks to Victoria Haire and Carol Betsch for refining, polishing, and improving the final manuscript and to George Whipple for the book's pleasing design. My editor, Holly Bailey, receives my deepest gratitude for her faith and skillful guidance.

I am also grateful to have received generous financial support. My initial work on Chapter 3 was supported both by the UCLA Center for International and Strategic Affairs and by the Chancellor's Office. The University of California's Institute on Global Conflict and Cooperation provided funds for preliminary work on Chapters 4 and 5, and an International Affairs Fellowship from the Council on Foreign Relations enhanced my sense of how policy makers actually make strategic calculations. My thanks also to UCLA's Academic Senate for its support of this project.

Finally, but foremost of those deserving acknowledgment, are my family. My mother has, as always, been an abiding source of encouragement and faith. My two little girls, Alexandra Tali and Joscelyn Ariella, are learning a great deal these days about conflict and cooperation. I love being with them, and I want them always to remain certain of their ability to effect change in their father's work habits—and in the world. My wife, Amy Elisabeth Davis, has been unconditionally supportive and delightfully demanding. She has been intellectual companion as well as best friend. She is the best of sounding boards, one who talks back. The most consistent imperative for writing this book was to dedicate it to her.

Arthur A. Stein

Encino, California

Why Nations Cooperate

1

Realism, Liberalism, and Dilemmas of Strategic Choice

In the village of Chelm the people argue. The moon, cry some, is more important than the sun. But others, fierce partisans of the sun, disagree. With the town rent by debate, the elders take up the question. After talking through the night, they decide: the moon is more important. It illuminates the otherwise dark hours. The sun, on the other hand, shines only in the day—when it is hardly needed.[1]

Conflict and cooperation both attend the workings of international politics. In academia the scholars argue. They disagree about which predominates, about which constitutes the norm from which deviations must be explained. Some see conflict as the hallmark of international politics and hold cooperation to be rare, of little consequence, and temporary. Others believe that international politics resembles other political systems in which there develop norms, rules, and a generally cooperative ambience. To them, conflict appears unusual. Scholars of both persuasions tend to concentrate their work on developing their presumptions about international politics and how these relate to patterns of either cooperation or conflict. Ironically, neither school focuses on explaining departures from the expected pattern. Rather, both schools emphasize what they perceive to be the norm.

Most basically, nations choose between cooperation and conflict, and

1. Other tales of the fictitious village of Chelm (pronounced with a gutteral *h*) can be found in Samuel Tenenbaum, *The Wise Men of Chelm* (1965; reprint New York: Collier Books, 1969).

such decisions underlie the entire range of international relations from alliances to war. When, how, and why they choose between them, and with what consequences, thus constitute the primary foci of the study of international politics.

In this book I examine the arguments of the two schools and argue that they are consistent with both cooperation and conflict—that the assumptions that underlie a conflict model of international politics are consistent with a great deal of cooperation and that the presumptions of cooperation are also consistent with conflict. I also elucidate the critical differences between the perspectives.

It is not surprising, of course, that international relations scholars do tend to concentrate on the extremes of conflict and cooperation, on war and alliances. Questions about their forms, causes, and consequences are critical. What brings nations into conflict? What leads nations to cooperate? When is cooperation institutionalized and formalized (as in alliances), and when is it less formal and less binding? These are some of the questions at the heart of the study of international politics and, therefore, of this book. In particular it asks why nations cooperate, a question of some importance for a world in which nations arm themselves in preparation for war and in which wars have occurred with some regularity.

Realism and Liberalism

Realism is the dominant intellectual perspective in the field.[2] Unlike utopianism and idealism, which focus on the world to be, realism connotes a hard-boiled willingness to see the world as it is—to accept extant reality unvarnished. Realists begin with a set of assumptions about international politics and emerge with a coherent perspective on international affairs.

Realists use anarchy as their primary metaphor for the international system. They stress that there exists no central authority capable of

2. My intention here is to characterize a large body of work that includes quite different and disparate strands. Moreover, as the realist literature has evolved, specific emphases and arguments have changed. Thus recent scholars have distinguished themselves as structural realists, neorealists, and modified structural realists. For discussions of realism and individual realists, see Kenneth W. Thompson, *Masters of International Thought: Major Twentieth-Century Theorists and the World Crisis* (Baton Rouge: Louisiana State University Press, 1980); James E. Dougherty and Robert L. Pfaltzgraff, Jr., *Contending Theories of International Relations: A Comprehensive Survey*, 2d ed. (New York: Harper and Row, 1981), chap. 3; and Michael Joseph Smith, *Realist Thought from Weber to Kissinger* (Baton Rouge: Louisiana State University Press, 1986).

creating and imposing order on the interactions of nation-states. Viewing countries as competitors in a state of nature, realists argue that the only order is that which emerges from competition under anarchy.

For realists, nation-states are the primary actors within this anarchic international system and nonstate actors, such as multinational corporations and domestic interest groups, are of secondary importance. Hence realists, unlike some others, treat world politics as international relations, that is, the relations between states. The international political arena is one in which states' policies clash.

In addition, realists view states as rational actors. They treat them as if they were individuals (the predominant label is unitary actor) who calculate costs and benefits and try to maximize their returns.[3] A nation sets its foreign policy as a rational response to a hostile and threatening international environment in which its survival can be ensured only by its own efforts. Irrespective of its domestic political system, its social and cultural traits, or the individual personalities of its leaders and citizens, a state is primarily and predominantly concerned with its own security.

Because the anarchic environment allows countries to expand without formal restraints, no individual nation's security can ever be ensured except through its own actions. Although states can pursue a variety of objectives, they must at least secure their continued survival. As rational actors, therefore, they focus on the means of providing security. Most fundamentally, they seek to maximize their own power. To realists, therefore, states in the anarchic world of international politics rely only on themselves.[4] They cannot tolerate intrusions on either their independence or their prerogatives. They must not allow themselves to become dependent on others. As a result, no division of labor or interdependence can be permitted to develop, especially between the great powers. Interdependent nations cannot by definition be great powers. Even lesser powers struggle to minimize their reliance on others.

---

3. A state can thus be treated analytically as a single unified and integrated entity in very much the same way that economists talk of firms.

4. The common characterization is to describe international politics as a system of self-help. This is a constant refrain in Kenneth N. Waltz, *Theory of International Politics* (Reading, Mass.: Addison-Wesley, 1979). The phrase comes from Frederick Sherwood Dunn, *Peaceful Change: A Study of International Procedures* (New York: Council on Foreign Relations, 1937), who also links self-help and power: "so long as the notion of self-help persists, the aim of maintaining the power position of the nation is paramount to all other considerations" (p. 13). Dunn is quoted in Kenneth N. Waltz, *Man, the State, and War: A Theoretical Analysis* (New York: Columbia University Press, 1959), p. 160.

Ever vigilant in this Hobbesian world of constant competition, struggle, and conflict, states continually prepare to defend themselves: "In
all times kings and persons of sovereign authority, because of their
independency, are in continual jealousies and in the state and posture
of gladiators, having their weapons pointing and their eyes fixed on
one another—that is, their forts, garrisons, and guns upon the frontiers
of their kingdoms, and continual spies upon their neighbours—which
is a posture of war."[5] Although the state of nature is not one of perpetual warfare, it is one of recurrent crises. It is, in short, a world of
perpetual conflict.

With no central authority to provide order, the only stability in this
dangerous and uncertain world comes from the competition itself. War
can be avoided only if the threat of war exists. And peace can be sustained only by preparation for hostilities, for in the realists' conflictual
world, peace is the absence of war. Cooperation is rare, because states
act autonomously and self-help is the rule. Since realists hold that
states cooperate only to deal with a common threat, they see cooperation, when manifest, as temporary or inconsequential and ultimately
explained by conflict.

In this vision international institutions are not particularly relevant.
States do not cede any authority to them, and they are powerless to
shape state behavior. Moreover, the cooperation essential to the func-
tioning of international institutions cannot exist.

This rejection of a role for international institutions has been a major
component of modern realism ever since it emerged in the late 1930s
as a self-conscious assault on the failure of the West to meet German
aggression. The realists portrayed themselves as hardheaded analysts
of the real world, one characterized by independent states prepared to
do anything to further their national interests. They contrasted themselves with utopians and idealists, whom they castigated for wishful
thinking. They ridiculed the interwar emphasis on international law
and international institutions, arguing that neither the League of Nations nor treaties to outlaw war could affect the fundamental nature of
international politics. The realists traced their intellectual lineage to
Machiavelli and Thucydides, whom they characterized as the first international relations theorist.

5. Thomas Hobbes, *Leviathan* (New York: Library of Liberal Arts, Bobbs-Merrill, 1958),
chap. 13, p. 108. For discussions of Hobbes and international politics, see Murray Forsyth,
"Thomas Hobbes and the External Relations of States," *British Journal of International
Studies* 5 (1979): 196–209; Hedley Bull, "Hobbes and the International Anarchy," *Social
Research* 48 (Winter 1981): 717–38; Gregory S. Kavka, "Hobbes's War of All against All,"
Ethics 93 (January 1983): 291–310.

In short, realists emphasize that states are autonomous and independent and concerned only with their own national interests. Moreover, they interact in an international environment in which there exists no overarching central authority to enforce order. This international anarchy leaves each state to fend for itself. In such a world, states expand until confronted and checked by others. Such a world is characterized by conflict and the constant possibility of war. Cooperation is unusual, fleeting, and temporary. International instititutions do not exist or are irrelevant.

In contrast to the realist vision lies a liberal one of a world in which self-interested actors engage in mutually rewarding exchange.[6] Rooted in nineteenth-century laissez-faire economics, liberalism argues that harmony and order emerge from such interactions between fully informed actors who recognize the costs of conflict.[7] Hence, self-interested rationality forms the basis of cooperation.

Although originally developed to explain the behavior of individual entrepreneurs and firms rather than world politics, liberalism contains a theory of international relations. For liberal arguments about cooperative exchange can be applied not only to companies but to other aggregate actors, including nations, as well. International trade theory, developed by liberal economists, treats states as the primary units and concludes that cooperative arrangements would emerge naturally from exchange. More generally, liberals hold that nations, wanting to maximize economic welfare, allow unfettered exchanges between themselves and other countries.[8] Since this exchange is based primarily on comparative advantage, it leads to a division of labor and to the growth of economic interdependence between states.

Liberals also see international interactions as akin to those that at-

6. The contrast drawn here is between realism and liberalism because both are proffered as positive, explanatory theories. Idealism and utopianism, on the other hand, are normative and concerned with creating alternative worlds. Like realism, liberalism is multifaceted, and what is or is not at its core can be disputed. For a fuller discussion of liberal arguments linking economic interdependence with international cooperation, see Arthur A. Stein, "Governments, Economic Interdependence, and International Cooperation," in Behavior, Society, and Nuclear War, vol. 3, ed. Philip E. Tetlock, Jo L. Husbands, Robert Jervis, Paul C. Stern, and Charles Tilly (New York: Oxford University Press, for the National Research Council of the National Academy of Sciences, forthcoming).

7. Ernest Gellner, "Trust, Cohesion, and the Social Order," in Trust: Making and Breaking Cooperative Relations, ed. Diego Gambetta (New York: Basil Blackwell, 1988), pp. 142–57, presents the Ibn Khaldunian view of social order, in which anarchy engenders trust or social cohesion, whereas government destroys it.

8. Indeed, the present challenge to international economics is to explain why trade barriers are so pervasive when formal deductions suggest that states should pursue free trade.

tend other social relations; they are characterized by the existence of rules, norms, and cooperative arrangements. The international system is no different from any other: it is characterized by regularity and order.

In addition, liberals draw an analogy between economics and international politics, between the order that characterizes markets and that which emerges from the self-interested behavior of states. Since liberals perceive the international system as comparable to a domestic market, they do not see the absence of an international government as preventing the emergence of cooperation. Given its roots in the economic liberalism of the late eighteenth and early nineteenth centuries, this view of the world is very much a laissez-faire one: order emerges as self-interested actors coexisting in an anarchic environment reach autonomous and independent decisions that lead to mutually desirable cooperative outcomes. Unlike realists, who stress the crises that attend the constant preparations for war, liberals point to peace as the norm. They see conflict as a periodic aberration that breaks the tranquillity in which exchange makes it possible for states to prosper.

According to liberals, conflicts arise out of misunderstanding and misperception. Only an inability truly to understand others, only hubris about the certainty of ultimate triumph, will result in conflict and war. With a little more understanding it will become apparent that the gains of conflict are illusory, and cooperation will become the inevitable result. Conflict reflects shortsightedness, miscalculation, misperception, or an absence of information.

Despite the different conclusions that they draw about the cooperative or conflictual nature of international politics, realism and liberalism share core assumptions.[9]

Although liberals avoid using the word "anarchy" to describe it, they share the realists' vision of the nature of the international system. This becomes, in fact, a critical justification for a specific disciplinary subconcentration on international politics as distinct from domestic politics; the distinction between anarchy and authority forms the basis for differentiating between foreign policy and other public policies. Although realists make the most of this point, liberals accept the obvious truth that there is no centrally mandated order in the international arena, that no hierarchical government exists to impose authoritative

---

9. I disagree here with those who suggest that realism and liberalism make different core assumptions. See, for example, Joseph M. Grieco, "Anarchy and the Limits of Cooperation: A Realist Critique of the Newest Liberal Institutionalism," *International Organization* 42 (Summer 1988): 485–507.

decisions on nation-states. Realists and liberals both recognize that there exists no system of global laws universally accepted as legitimate and binding and enforceable by a central administration with power and authority.

Although liberals recognize the absence of a central authority in the international system, they reject the realists' metaphor, for anarchy connotes chaos and conflict. Liberals disagree with the presumptive consequences that realists see as emerging from the absence of central authority. John Mueller, for example, suggests replacing anarchy with the characterization "unregulated."[10] Hedley Bull accepts the metaphor but signals his disagreement about the consequences involved by titling his book *The Anarchical Society*. The international system may indeed be anarchical, he admits, but it remains a society.[11]

In addition, realists and liberals both view states as relevant actors in world politics. Owing to its emphasis on this formulation, realism is often dubbed a states-as-actors model of international politics. Liberalism, on the other hand, focuses not exclusively on nation-states but also on individuals and firms.[12] Indeed, because the late eighteenth-century liberals juxtaposed the public policies they presumed would emerge from representative governments with those pursued by monarchies, liberalism is often characterized as a perspective that reduces international relations to domestic politics.[13] But liberalism at its core focuses on actors, whether individuals or collections of individuals, and on the results of interaction between self-interested actors. Thus, although liberals may argue for the importance of actors other than nation-states, they readily recognize, and their arguments can and should be readily applicable to, states as actors.[14]

10. John Mueller, "Realist Theory and Practice," presentation to the Workshop on International Strategy, University of California, Los Angeles, October 5, 1989.

11. Hedley Bull, *The Anarchical Society: A Study of Order in World Politics* (New York: Columbia University Press, 1977).

12. Most recently, liberals have focused on the importance of transnational actors; see Samuel H. Huntington, "Transnational Organizations in World Politics," *World Politics* 25 (April 1973): 333–68.

13. In Waltz's terms, liberalism is a second-image argument; see Waltz, *Man, the State, and War*. Stephen D. Krasner, in *Defending the National Interest: Raw Materials Investments and U.S. Foreign Policy* (Princeton, N.J.: Princeton University Press, 1978), equates liberalism with a pluralist argument of foreign policy. For my discussion of the early nineteenth-century liberals' views of the state and international cooperation, and how these evolved, see Stein, "Governments, Economic Interdependence, and International Cooperation."

14. Those who believe that nonstate actors are important typically argue that these other actors matter in the ways in which they constrain or affect state choices. It is still states that choose to go to war or to enter alliances. State policy matters but can be affected by other actors as well.

Finally, both realism and liberalism presume self-interested, purposive, and calculated behavior. In its strict form, the rational-actor model presumes that actors have complete information, assess all options, and then maximize some hierarchy of values. But an argument need not meet all these requirements in order to retain its character as a *purposive-actor explanation*. A constrained rationality argument can be used to explain behavior that is deemed purposive and based on a choice that reflects calculation—if, that is, options are selected with the expectation that they will provide better rather than worse outcomes. In realism, the presumption appears in the formulation that states respond rationally to the challenges posed by the anarchic environment in which they must compete and struggle. In liberalism, it comes with the view that actors rationally pursue their self-interest.[15]

The similarity between modern realism and liberalism is evinced by the connections both perspectives have to the ideas and methods of economics. This is most obvious in the case of liberalism, which has historic roots in the works of such figures as Adam Smith and John Stuart Mill. Although they were known in their own time as political economists, their thought has played an important role in the development of both modern economics and political science. Modern realism also borrows essential ideas from economics. In his two main works Kenneth Waltz, who is currently the key intellectual figure in realist thought, conceptualizes the international system as an anarchic world populated by competing self-interested states; the view is clearly linked to the notion of an economic market with competing firms.[16] In Waltz's words, "Balance-of-power theory is microtheory precisely in the economist's sense. The system, like a market in economics, is made by the actions and interactions of its units, and the theory is based on assumptions about their behavior."[17]

15. As mentioned in earlier footnotes, not all realists and liberals fit all my characterizations of them. Reinhold Niebuhr, for example, an important realist figure, castigated liberals for their emphasis on self-interest, arguing that it blinded them to "the irreducible irrationality of human behavior." Quoted in Thompson, *Masters of International Thought*, pp. 21–22.

16. Waltz's position as the central figure of modern realists is reflected by his work's central position in an edited volume, *Neorealism and Its Critics*, ed. Robert O. Keohane (New York: Columbia University Press, 1986). This volume, in addition to republishing four of nine chapters from Waltz's *Theory of International Politics*, contains several critical essays and concludes with Waltz's reply to his critics. That Waltz constantly uses economic metaphors can be seen in his two volumes on international relations theory, *Man, the State, and War* and *Theory of International Politics*.

17. Waltz, *Theory of International Politics*, p. 118. He compares nations in an international system to oligopolistic firms (p. 105, for example), and he seems to have been affected by economists' studies of imperfect competition. Richard Rosecrance criticizes

Indeed, the realists' very distinction between different international systems as unipolar, bipolar, and multipolar is drawn from economics.[18] In each case the world is viewed as anarchic, and nation-states are seen as acting autonomously, but the number of great powers in the system generates different patterns of conflict and interaction, and these carry different consequences for systemic stability.[19] This categorization mirrors exactly the economists' differentiation between monopolistic, duopolistic, oligopolistic, and competitive markets.[20] These different economic structures are characterized by different degrees of competition, with resultant consequences for prices.

Realists borrow heavily from the methods of economics as well. Game theory, widely used to model economic behavior, quickly came to be seen as a way to model international phenomena.[21] In fact, it became the basis for important contributions by economists to the study of international politics, especially in the area of military strategy.[22] Certain games, especially the prisoners' dilemma and chicken, have been widely used as generic metaphors for international phenomena.[23]

Waltz for ignoring international trade theory and thus missing the cooperation and interdependence in the international system. See Rosecrance, "International Theory Revisited," *International Organization* 35 (Autumn 1981): 691–713.

18. Some scholars do not include unipolarity; others include hybrids. Morton A. Kaplan, for example, discusses six types: balance of power, tight bipolar, loose bipolar, universal, hierarchical, and unit veto. See Kaplan, *System and Process in International Politics* (New York: John Wiley, 1957).

19. The polarity literature is reviewed by R. N. Rosecrance, "Bipolarity, Multipolarity, and the Future," *Journal of Conflict Resolution* 10 (September 1966): 314–27.

20. The isomorphism between the typologies used by international relations theorists and economists is noted by Kenneth Boulding in his review of Morton Kaplan's book. See Boulding, "Theoretical Systems and Political Realities: A Review of Morton A. Kaplan, *System and Process in International Politics*," *Journal of Conflict Resolution* 2 (December 1958): 329–34.

21. The classic work that created the field is by John von Neumann and Oskar Morgenstern, *Theory of Games and Economic Behavior* (Princeton, N.J.: Princeton University Press, 1944). In a discussion of game theory in *Man, the State, and War*, pp. 201–5, Waltz demurs that "the reference to game theory does not imply that there is available a technique by which international politics can be approached mathematically" (note, p. 201). But he continues to hold that international politics can be "profitably" described using game-theoretic concepts.

22. The most famous example is that of Thomas C. Schelling, *The Strategy of Conflict* (Cambridge, Mass.: Harvard University Press, 1960).

23. See, for example, the way in which games are used as metaphors for key arguments in international politics in Glenn H. Snyder, " 'Prisoner's Dilemma' and 'Chicken' Models in International Politics," *International Studies Quarterly* 15 (March 1971): 66–103, and Robert Jervis, *Perception and Misperception in International Politics* (Princeton, N.J.: Princeton University Press, 1976). For benchmarks in the use of game theory in the study of international relations, see, in addition to Schelling's *Strategy of Conflict* and Snyder's article, Richard E. Quandt, "On the Use of Game Models in Theories of International

Yet despite their common focus on self-interested states interacting
in an anarchic environment, realists and liberals come to different conclusions about the nature of international politics. Realists see a world
of conflict in which cooperation is fleeting, and tension the norm. Conflict, they argue, is rooted in the very nature of international politics, in
the constant struggle for power and survival that characterizes a world
of autonomous independent actors making self-interested choices.[24]
Liberals, on the other hand, see autonomous self-interested behavior
as consistent with the emergence of order and cooperation. Perceiving
a laissez-faire and cooperative world, they understand conflict to be
wasteful, destructive, and inefficient. Actors arrive at mutually advantageous arrangements that sometimes involve the development of overarching institutions. To liberals, therefore, conflict must be a product
of imperfect or incomplete information. Since conflict grows from miscalculation and misperception, it can be avoided.

International relations involve both cooperation and conflict, evinc-
ing more cooperation than realists admit and more conflict than liberals recognize. In this book I assess the bases of cooperation and
conflict and the implications of realist and liberal premises. I focus
especially on the former—on the conditions necessary for cooperation
to emerge in an anarchic world and on the forms that cooperation
takes. In doing so, I qualify both realist and liberal arguments and
develop adjuncts to both. I demonstrate that realist assumptions are
consistent with international cooperation and liberal assumptions with
international conflict.

Indeed, I make clear the ways in which realists and liberals are correct and the extent to which they are not. Both schools of thought
correctly describe behavior that occurs within limited domains, under
particular sets of circumstances, and given actors' specific calculations

Relations," *World Politics* 14 (October 1961): 69–78; Duncan Snidal, "The Game *Theory* of
International Politics," *World Politics* 38 (October 1985): 25–57; and Robert Jervis, "Realism,
Game Theory, and Cooperation," *World Politics* 40 (April 1988): 317–49. For recent book-
length treatments, see Steven J. Brams, *Superpower Games: Applying Game Theory to
Superpower Conflict* (New Haven, Conn.: Yale University Press, 1985), and Steven J. Brams
and D. Marc Kilgour, *Game Theory and National Security* (New York: Basil Blackwell, 1988).
But see especially the new book by Robert Powell, *Nuclear Deterrence Theory: The Search
for Credibility* (Cambridge: Cambridge University Press, 1990). For a recent comprehensive
introduction that relates game theory to political science in general, see Peter C. Ordeshook, *Game Theory and Political Theory: An Introduction* (New York: Cambridge University Press, 1986).
    24. In Waltz's words, "competition and conflict among states stem directly from the
twin facts of life under conditions of anarchy"; see his "The Origins of War in Neorealist
Theory," *Journal of Interdisciplinary History* 18 (Spring 1988): 615–28, quote from p. 619.

and assessments of self-interest. I also delineate the core differences from which the conflicting deductions of realists and liberals derive. Specifically, I pinpoint their most critical disagreement in their assumptions about the decision criteria that states use in calculating their interests. Yet I argue that both sides are correct. In different circumstances, different bases of calculation are appropriate. Cooperation and conflict, I argue, result from the forces of circumstance, from a set of situational factors that I delineate.

Strategic Interaction

This work begins with the common assumptions of realism and liberalism and focuses on the decisions that nations make about whether to cooperate with one another. Most fundamentally, my analysis rests on the twin premises of the phrase "international relations." I treat nations as the salient actors and presume that the proper focus of study is relations. These assumptions necessitate a focus, therefore, on the dynamics of cooperation and conflict between nation-states. Like marriage, wars and alliances presuppose the existence of two interacting parties.

Indeed, not all nations interact. It is possible to analyze the behavior of European nations in the era before they discovered the New World without referring to the Indian nations of the North American continent. The reason, of course, is that the former did not know of the latter's existence, and vice versa. Isolation, the absence of contact and interaction, was for most of world history a not uncommon phenomenon.

Over time, however, the evolution of science and technology has eroded that isolation. The advent of steamships and the development of air travel meant that people could travel from one part of the world to another with increasing speed. Advances in technology have also altered the ways in which information flows, allowing the far reaches of the world to be linked instantaneously.

In short, the emergence of a single global system and international society has brought about a vast increase in the number of international interactions and has ended the isolation of countries. Areas whose former development could be assessed without reference to a wider world came to be incorporated into one system.[25] To some degree and

25. Hedley Bull and Adam Watson, eds., *The Expansion of International Society* (New York: Oxford University Press, 1984); Immanuel Wallerstein, *The Modern World-System: Capitalist Agriculture and the Origins of the European World-Economy in the Sixteenth*

in some form, every nation interacts with every other nation, and since
it takes at least two nations to fight a war or to form an alliance, explaining the occurrence of either phenomenon requires elucidating the
behavior of all participants and how their actions lead to a particular
event.

The study of international relations generally focuses on or begins
with dyadic ties, for the bilateral link represents the most basic form
of relationship.[26] Moreover, delineating the workings of even simple
two-country links provides a critical foundation for understanding the
more complex world of international politics within which they exist.[27]

Further, this study presumes states to be independent decision makers, a conceptualization rooted in the classic realist characterization of
international politics as relations between autonomous countries dedicated to self-preservation, ultimately able to depend only on themselves, and prepared to resort to force in an anarchic world. In this
view, nations consider every option available to them and make their
choices independently in order to maximize their own returns. As
Waltz puts it, they "develop their own strategies, chart their own
courses, make their own decisions."[28]

This independence lies at the heart of the metaphor of international
politics as anarchy and forms the basis for treating states as equals.[29]
For the national autonomy provided by sovereignty is different from
the hierarchical relationships that exist within a domestic political
system.[30]

That states are equal does not mean that they are the same across
all dimensions. They may be large or small, culturally homogeneous or
heterogeneous, surrounded by water or by other countries. Their public
policies may derive from the workings of democratic political institutions or reflect the preferences of individual dictators. States are the
same only in that they are all sovereign.

Moreover, the autonomy and sovereignty of nation-states do not

---

Century (New York: Academic Press, 1974); Wallerstein, *The Modern World-System II:
Mercantilism and the Consolidation of the European World-Economy, 1600–1750* (New
York: Academic Press, 1980); and work on the evolution of world-system capitalism.

26. In fact, an even more basic form of relationship—that between an actor and nature—does exist.

27. This dyadic approach is used to analyze not only other international issues, such
as trade, but also such other strategic interactions as interpersonal relations.

28. Waltz, *Theory of International Politics*, p. 96.

29. Robert A. Klein, *Sovereign Equality among States: The History of an Idea* (Toronto:
University of Toronto Press, 1974).

30. See the discussion by Waltz, *Theory of International Politics*, pp. 88–99.

mean that they always remain unaffected by the preferences and actions of other countries. The existence of sovereignty implies only that
a state decides for itself, not that others cannot circumscribe its choices.
The ability to decide for itself does not prevent others from defining its
options as a rock and a hard place. A state can be independent in the
sense of having decisional autonomy and yet be fully dependent in the
sense that its decisions are inconsequential. In other words, independence and dependence can be used to describe the locus of control over
outcomes and to characterize relationships as well as to refer to the
legal status of political autonomy. To be independent in a relational
sense means not to rely on another for outcomes. Whatever one gets
is obtained independent of the actions of others.

As the leaders of many newly born nations have quickly discovered,
for example, legal independence can coexist with enormous dependence on former colonial masters. Official colonialism has often been
replaced, in fact, by an unofficial neocolonialism. Although the legal
status of sovereign nation provides only the autonomy to make decisions, the harsh realities of the world are such that every former colony,
like young adults who leave their parents' homes to strike out on their
own, must make its way in a world of constraints. Freedom does not
imply the availability of all options. Sovereignty means only that a nation must rely on itself for its own survival and can choose for itself
given its circumstances and options. Moreover, sovereignty comes with
no guarantees: nations can, and do, disappear.[31]

A state may be independent, have its own flag, anthem, army, and
political system and yet remain dependent in the sense that its fate is
not in its own hands as much as in others'. An autonomous sovereign
state is free to make its own choices and has its own mechanisms and
procedures for arriving at decisions. Those decisions are not dictated
to it by others. Yet having this independent decision-making ability
does not mean that a state cannot have its payoffs determined by
others. This, of course, is precisely what many third-world nations discovered when they received their independence. No longer colonies
unable to exercise independent judgment because their choices were
dictated to them, they became able to make independent and autonomous decisions. Yet this change of status did not mean that they
obtained the ability to affect their own returns. In many domains they

31. See J. David Singer and Melvin Small, *The Wages of War, 1816–1965: A Statistical
Handbook* (New York: John Wiley, 1972), pp. 19–30. In 1948 the end of the British mandate
for Palestine and the declaration of Israeli nationhood was the occasion for the start of
an Arab invasion. In short, sovereignty is not ensured and must be continually secured.

remained dependent, for others' decisions fully determined their payoffs.[32] The existence of choice and the freedom to choose are not necessarily consequential for one's payoffs.

A focus on strategic interaction, and the common emphasis on self-
interest in realist and liberal arguments, presumes the existence of
choice, and the least degree of choice is between two options. Given
their emphasis on cooperation and conflict, international relations theorists, like social analysts more generally, crudely dichotomize the strategies available to states as cooperation and defection. They begin with
a world of two nations, both independent decision makers needing to
choose between two options. These nations' combined choices generate four possible outcomes, each of which carries a payoff for each
actor.

States choose strategies, not outcomes. This important point is the
source of much confusion. In Figure 1, for example, two actors, A and
B, choose between cooperation and defection. The cell entries show
how each actor ranks the four possible outcomes. Each actor most
prefers to defect and have the other cooperate; each least prefers to
cooperate and see the other defect. One's second choice is mutual
defection; the other's second choice is mutual cooperation. Each actor
has a dominant strategy, a course of action it prefers irrespective of the
other's decision. Both want to defect if the other cooperates, and both
want to defect if the other defects. The outcome, obviously, is one of
mutual defection. Let us presume that defection in this example constitutes hostile escalation during a crisis and that mutual defection
represents war. Clearly, actors' autonomously chosen strategies lead to
war. This does not mean, however, that war is in either's "interest."
Each most prefers outcomes other than war. But since both want to
escalate if the other does not, and because neither wants to capitulate
while the other escalates, war results from their individual strategic
choices. Neither wants nor intends war: each actor most wants the
other to capitulate. Yet each knows that because it will defect regardless
of what the other does, the outcome of mutual defection is a real possibility. In other words, each understands that its own course of action
entails the chance, if not the likelihood, of war.

---

32. Not surprisingly, this led to the revival of arguments about neocolonialism and
neo-imperialism and dependency. "Dependence" is also used here quite differently from
its meaning in the international political economy literature. For a discussion of the
terminology in that debate, see James A. Caporaso, "Dependence, Dependency, and
Power in the Global System: A Structural and Behavioral Analysis," *International Organization* 32 (Winter 1978): 13–43.

Figure 1. A situation of pure conflict

| | Actor B | |
	Cooperate	Defect*
Cooperate	2, 3	1, 4
Actor A		
Defect*	4, 1	3, 2**

In this and following figures, cell numerals refer to ordinally ranked preferences: 4 = best, 1 = worst. The first number in each cell refers to A's preference, and the second number in each cell refers to B's preference.

*Actor's dominant strategy
**Equilibrium outcome

Again, actors have preferences among outcomes and their associated payoffs. But actors do not choose outcomes; they choose strategies, and outcomes result from the combination of actors' decisions. As fictional British Cabinet Secretary Sir Humphrey Appleby said to Prime Minister James Hacker, "Things don't just happen because prime ministers are very keen on them. Neville Chamberlain was very keen on peace."[33]

The constellation of actors' preferences underlies state strategies and resultant outcomes. If states have the interests given in Figure 1, for example, they will conflict. We assume that both states are aware of their interests and are rational and that outcomes are determined by the structure of their preferences. The interests of the states in Figure 1 are diametrically opposed. In this case, moving from cell to cell improves one actor's lot and worsens the other's. In fact, each actor's best outcome is the other's worst. Each one's choice is clear, as is the outcome. Assuming rationality and knowing the actors' preference orderings allow a complete explanation of choice and outcome.

The interests of the states in Figure 2, on the other hand, are largely consonant. They most prefer the same outcome, and each has a dominant strategy not only for itself but for the other as well. That is, actor A not only prefers A_1 regardless of what actor B does but wants actor B to do B_1 regardless of A's behavior. B_1 is in fact B's dominant strategy. And B always wants A to do A_1. In other words, each wants to do what the other wants it to. Individual rationality leads to a mutually desired outcome. Knowing only the structure of the preferences and presuming the rationality of the actors are enough to explain choice and outcome.

33. Television series "Yes, Prime Minister," episode "The Ministerial Broadcast."

Figure 2. A no-conflict situation

Actor B

	B$_1$*	B$_2$
A$_1$*	4, 4**	3, 2
A$_2$	2, 3	1, 1

Actor A

*Actor's dominant strategy
**Equilibrium outcome

To explain the determination of the cooperative and conflictual outcomes delineated in Figures 1 and 2 does not require any analysis of
strategy or interaction. The outcome derives from the structure of the
payoff environment, and thus cooperation and conflict are situationally
or structurally given. The only interesting questions left for scholars to
ask involve the origins or bases of the actors' particular preferences.
The constellation of actors' preferences in any situation is one characteristic of their relationship.

The debate between liberals and realists can be conceptualized as
being between scholars who see each of the above figures as the modal
payoff environment in international politics. If realists argue that Figure
1 captures international relationships, conflict is a given. One merely
need amend the above realist assumptions to include conflictual preferences as a given.[34] Similarly, if liberals argue that international relations are captured by Figure 2, cooperation is a given.[35] Again, one need
merely amend the liberal assumptions discussed above to include cooperative preferences. But then cooperative and conflictual preferences
would be part of the assumptions of liberalism and realism, and their
deductions about international behavior would not be compelling. For
their debate would revolve not around deductions but around underlying assumptions.[36] And there is no reason a priori to argue that only

34. This is essentially what Robert Gilpin does when he writes that "all realist writers
...share three assumptions....The first is the essentially conflictual nature of international affairs"; see Gilpin, "The Richness of the Tradition of Political Realism," *International Organization* 38 (Spring 1984): 287–304, quote from p. 290.

35. E. H. Carr argues that the classic liberal arguments on behalf of laissez-faire assume
a harmony of interests; see Carr, *The Twenty Years' Crisis, 1919–1939: An Introduction to
the Study of International Relations* (New York: St. Martin's Press, 1939).

36. Contradictory assessments of human nature can also be said to lie at the root of
the different conclusions drawn by liberals and realists. Liberals emphasize human per-
fectability; realists focus on inherent human aggressiveness.

particular sets of preferences characterize international relations. If this is to be derived, it cannot be assumed. And if the situations captured in Figures 1 and 2 are both possible, then, so far at least, cooperation and conflict are both possible.

The foregoing discussion uses quite simple game theory, which captures the elements of interaction, autonomous decision making, choice, cooperation, and conflict that are the components of relations between nations. Because this book focuses on the strategic interactions of states and the bases of cooperation and conflict between them, however, it should be understood as a study of international relations. It is no more intended as a contribution to game theory than a regression analysis of wars is intended to be a contribution to econometrics.[37]

Moreover, despite my inclusion of some simple original deductions relating to international relations, I do not pretend to develop new formalizations and new equilibrium concepts. Nor have I written a guide to new advances in game theory. Although I cite some recent work, I do not apply or use all of modern game theory. Footnotes are provided for those interested in learning how more sophisticated recent work reinforces conclusions I reach here by quite simple means.

In addition, I am fully prepared to relax, and even challenge, the assumptions typically made in formal game-theoretic work. My concern here is the analysis of strategic international interaction. My goal is to illuminate our understanding of the dynamics of international cooperation and conflict. I have no interest in using artificial constructions of international relations to illustrate arguments developed by mathematicians and economists. I believe, in short, that the overriding concern of scholars should be to make their models isomorphic with the reality they wish to analyze.

This work is specifically addressed to those interested in international relations. Every effort has been made to make the manuscript as accessible as possible and not to assume any technical expertise on the part of the reader. The focus is on bottom-line conclusions rather than on technical pyrotechnics. Those with a more mathematical predisposition are encouraged to bear with the simple exposition and concentrate on the ideas and arguments.

Focusing on strategic interaction in this fashion allows a conceptualization of international relations as involving both cooperative and conflictual strategies that emerge from a set of circumstances and cal-

37. I find "game theory" an unfortunate appellation and prefer the phrase "models of strategic interaction."

culations. Such an approach integrates both phenomena in a single perspective and does not necessitate presuming at the outset that international relations are either inherently conflictual or intrinsically cooperative.[38] It allows for an analysis of international institutions and arms races on the same bases and using a single framework. In other words, it can explain both norms of cooperation and unbridled rivalries. Moreover, in emphasizing cooperation and conflict and their coexistence, the book provides a counterweight to older realist views that equate an anarchic international order with conflict and to more recent liberal analyses that see the international system as a marketplace riddled with cooperation. Because a strategic-interaction approach provides a framework for the study of both security and economic issues, it allows analysts to use concepts, theories, and arguments from the subfields of both national security policy and international political economy, and illustrations from these different substantive domains appear in the ensuing chapters.

Dilemmas and Paradoxes

This book also emphasizes that choices are not always clear-cut. Confronted by the need to choose, nations, like individuals, must weigh the costs and benefits that seem likely to attend each path. Whether the issues are economic or military, they must pick the strategy that seems best to serve their interests. Sometimes, of course, the choice is straightforward, for the payoffs that result from the different options clearly dictate a specific course for rational self-interested actors. In words that will be clearer later, the payoff environment, conjoined with assumptions of rationality and full information, leads to an unambiguous choice. In other situations, however, states find rational logics for competing options. As the existence of domestic debates over foreign policy suggest, either countries are no more than agglomerations of individuals and groups, or, at the very least, there are multiple assessments of national interest and strategy.

This book is about the choices that nations make and the dilemmas presented by the need to choose when no option presents a singularly

38. This analytic tendency to bifurcate the field is exacerbated by prevalent methodologies for empirical work. Scholars have an unfortunate tendency to select comparative case studies on the basis of a truncated dependent variable. To give one example, a set of cases of deterrence failures is chosen to see what they have in common. Such an empirical procedure not only is methodologically flawed but also entails a separation of the study of conflictual and cooperative behavior.

powerful case. It is about the logic of picking a strategy when there are rationales and rationalities for competing and conflicting choices. It finds both domestic debates about foreign policy and scholarly debates about international cooperation to be rooted in competing rationalities and particular circumstances.

The existence of rationales for competing options generates dilemmas of choice and, therefore, paradoxes. A classic individual dilemma is that of choosing between what is best for oneself and what is best for one's group or collectivity. Paradoxically, in pursuing what is individually rational, an actor makes itself and others worse off than had it pursued what was collectively rational. Alternatively, intertemporal dilemmas arise when states find themselves forced to choose between short-term and long-term interests. Immediate payoffs can be maximized at the expense of future ones, or prospective payoffs can be maximized through current self-denial and sacrifice.

General dilemmas have specific substantive manifestations in international politics. The "security dilemma," which sometimes arises for makers of national security policy, provides one example. Although states arm themselves to ensure their security, one possible consequence is that others will respond in kind; in other words, the desire for safety can lead to actions that result in greater insecurity. The failure to arm has its own potentially dire consequences, however. In short, strong imperatives exist for both options. It makes sense for a nation to forgo arming itself in the hope that a world without weapons will result in greater security. Yet it also makes sense to deploy arms in order minimally to ensure that others cannot take advantage of an undefended state.

States confront economic dilemmas as well. They can cooperate and pursue liberal economic policies that improve global welfare and efficiency. Yet in some cases a nation can improve its relative economic position by defecting and cheating on others. Each course of action makes sense. Yet one consequence of a state's decision to maximize global welfare can be that it undercuts its own relative economic position. On the other hand, maintaining one's relative position may mean the loss of potential absolute wealth.

In this book I investigate the ways in which the decision to cooperate or conflict derives from the interaction of the payoff environment, actors' perceptions, their views of intertemporal trade-offs, and their relationships with others in dilemmas of strategic choice. The particular dilemmas of strategic choice analyzed here presume interdependence between states. They arise because states' choices affect both themselves and others. Thus the ability to analyze the decision to cooperate or conflict when powerful logics exist for each requires the use of a strategic-interaction approach, one that focuses on choices and their consequences in a context of interaction.

In dilemmas of strategic choice, structure is important but not fully determinative. I consider here situations in which structure does not completely determine choice but in which indeterminacy makes strategy and bargaining important. This approach contrasts with studies that posit the nature of interaction and derive outcomes from antecedent context, that view structure as determining the constellation of preferences among which states choose, and that see the nature of state interaction as fully derivable.

Individual chapters address the circumstances that underlie cooperation, assessing both liberal and realist arguments in the process. Chapters 2 and 3 demonstrate that either cooperation or conflict can result from autonomous self-interested behavior and from misperception. Realist assumptions can in some circumstances generate the cooperation more typically predicted by liberals. Similarly, the validity of liberal arguments is constrained by circumstance. Misperception can be the basis, depending on context, both for otherwise avoidable conflict and for otherwise avoidable cooperation. Both chapters take as given that states are self-interested in a narrow individualistic sense. The strategies discussed in them deal with departures from autonomous rationality because of either misperception or a desire to improve one's outcomes.

More specifically, Chapter 2 demonstrates that realist assumptions can generate different kinds of international cooperation. For whenever actors can improve on the outcomes that would arise from their individually rational behavior, they have an interest in eschewing autonomous decision making in favor of international regimes to ensure coordination and collaboration. In other words, realist assumptions can be consistent with states departing from those courses of autonomous action they would follow in order to obtain more desirable outcomes. But such departures from autonomous individualistic behavior arise only in particular circumstances.

In contrast, Chapter 3 tackles a central component of the liberal argument, the role of misperception. It focuses on the implications of misperception for state behavior and for international cooperation and conflict. There are times when a nation's decision whether or not to cooperate depends on its expectations of what others will do. In such cases, misperception can affect its actions, leading, as liberals suggest, to otherwise avoidable conflict. But it also shows that misperception can also lead to otherwise avoidable cooperation. The implications of misperception necessarily depend on circumstance.

The findings of Chapters 2 and 3 derive from situations with varying payoffs. They show that individualistic decisions about whether or not to cooperate are determined by the payoff environment and that self-interested calculations can generate dilemmas for nations.

In Chapters 4, 5, and 6, however, I analyze another set of dilemmas, ones that arise not from the payoff environment but because different decision criteria lead to different courses of action. In developing arguments about departures from narrow self-interest, I focus here on alternative conceptualizations of rationality and the circumstances in which there are competing logics for choice. I deal with dilemmas generated by a conflict between various ways of assessing self-interest.

Chapter 4 details the implications for international politics of the realist emphasis on survival. In the international arena, states fear for their lives. Nations do disappear, and there is no guarantee of continued existence in the anarchic international arena. States whose survival is not guaranteed can be confronted with a choice between ensuring security and maximizing gains. Moreover, there are asymmetries in international politics. Some nations are more powerful than others. Some gain more (or lose more) than others in international relations. As a result, some can look with greater equanimity than others at potential losses, and some can afford to think of long-term returns. In Chapter 4 I discuss the implications of states' different temporal horizons and different status quo points. I show that states are sometimes confronted with dilemmas between long- and short-term views and between ensuring security and maximizing payoffs. I also develop the implications of the possibility of national extinction and of asymmetries of power and payoffs for international cooperation. In certain situations nations must make a choice between different criteria for assessing and calculating self-interest.

Chapter 5 addresses the implications of competition for international cooperation and conflict. Both realists and liberals emphasize competition, but they mean quite different things by it. The realist emphasis on power implicitly stresses the importance of relative calculations, whereas the liberal focus on wealth typically implies a concern with absolute returns. In Chapter 5 I outline the reasons for and the consequences of a nation's decision to pursue relative advantage rather than maximize absolute outcomes. I maintain that a variety of international interactions are inherently competitive and adduce the consequences of international struggles for advantage. In the final analysis, I argue, the existence of competitive calculations, and not merely the absence of a global central authority, lies at the root of many international conflicts.

In contrast, Chapter 6 focuses on especially cooperative relationships, ones in which states attach some utility to the consequences of their actions for other nations. Here I demonstrate that the international system contains more cooperative behavior than is suggested by either realists or liberals, and show that states will in some cases make choices on the basis of their allies' concerns. The very competitive underpinnings of some international relations, I argue, lead to the emergence of true helping behavior that cannot be fully explained with reference to individualistic national interest.

Along with building general and abstract arguments about the bases of international cooperation and conflict, I delineate my analyses substantively, drawing illustrations from diplomatic history, from studies of the international political economy, and from studies of national security policy. Among my circumstantial arguments is a bleak assessment of the prospects for arms control, an analysis derived from an explication of the competitive underpinnings of arms races and their not being amenable to negotiated solutions. A role does exist for arms control in resolving arms races, however, when one state is prepared to accept a relatively inferior position. More optimistically, I provide a generally positive assessment of the prospects for maintaining a global economic order in the wake of a relative decline in American economic power. This view stems from an examination of the strategic bases of cooperative regimes.

Ironically, the most hopeful and most pessimistic conclusions in this book derive from the same basic insight: conflict is ever possible in international politics, but even some of the strongest forms of cooperation sometimes depend on this possibility. Indeed, the outcomes of dilemmas, of nations' needing to choose between cooperation and conflict when compelling logics exist for both, all too often balance on a knife's edge.

2

Coordination and Collaboration: Dilemmas of Common Interests and Common Aversions

For realists and liberals alike, the need to explain the existence of order and international institutions poses a problem. Continuous and consistent efforts have been under way for more than one hundred years to develop international regimes, more or less institutionalized arrangements for structuring international relationships in various domains. Such regimes have provided a patchwork quilt of international rules and international institutions in the anarchic world of international politics. International relations theory must, therefore, find a way to explain order amid anarchy.

Realists hold that since sovereign nations act autonomously in their own self-interest, international institutions are inherently irrelevant to world politics. Their existence does not change state behavior. Realists cannot really explain the existence of regimes or even the effort of states to create them.

Liberals, on the other hand, believe there should be no need for such arrangements. The self-interested behavior of actors in a competitive setting should result in mutually beneficial exchanges that require no international regimes. Like a well-functioning market in which self-interested behavior leads to optimal, efficient outcomes, an anarchic international system composed of self-interested states should need no regulation.

Moreover, since both liberals and realists emphasize self-interest and the absence of any international arrangements that do more than reflect the immediate self-interest of nation-states, both find it difficult to ex-

plain why countries would create such institutions. Both also find it
hard to account for the roles of such regimes once created. After all,
institutions that comport with actors' interests are unnecessary, and
any that deviated from those concerns would not be efficacious, for the
states would merely pursue their interests.[1]

Hence it is not surprising that many scholars define "international
regimes" so broadly as to constitute either all international relations or
all international interactions within a given issue area. In this sense an
international monetary regime is nothing more than all international
relations involving money. Such use of the term "regime" does no more
than signify a disaggregated issue-area approach to the study of international relations, and, so defined, "regimes" have no conceptual status; they do not circumscribe normal patterns of international behavior.
They do no more than delimit the issue domain under discussion.
Similarly, a conceptual definition of "regimes," as, for example, "the
rules of the game," in no way limits the range of international interactions to which it refers. We can, after all, describe even the most
anarchic behavior in the international system as guided by the rules of
self-interest or self-help.[2] To specify the rules of the international po-

1. An alternative view takes institutions as given and sees their role as sociological
and formative. Institutions represent structures that socialize actors and shape their
interests. Such an approach is quite important in understanding individuals within society. When studying domestic politics, where we are able only to surmise the origins of
states, we can still recognize that they transcend and shape individuals. States, local
communities, ethnic groups, and religious organizations all give form to the worldviews
and interests of people. They transmit rules, norms, expectations, and obligations. Individuals and their interests are products of nature but also of nurture; their views and
choices are partly determined by the institutions that shape and mold them. Such a
sociological view of institutions cannot be sustained for the international system. Not
only do states transcend most international institutions, but the ways in which international structure peculiarly socializes nations is quite unspecified. Moreover, the institutions that constitute structure for individuals can be readily delineated, whereas the
constitutive structures for states are amorphous and not substantiated. Finally, although
one can imbed any agent within some larger structure, those that are human artifacts
must somewhere be products of agency. This so-called agent-structure problem, imported into international relations from sociology, is of greater interest to the study of
domestic politics, where the socializing role of institutions is greater. The key issue in
international relations is explaining how international regimes are products of agency.

2. This is the basis of a disagreement I have with the views of several other scholars.
Donald J. Puchala and Raymond F. Hopkins, for example, treat international regimes as
coextensive with international politics; see Puchala and Hopkins, "International Regimes:
Lessons from Inductive Analysis," *International Organization* 36 (Spring 1982): 245–75.
Similarly, although Oran Young does not formally equate international politics with regimes, his definitions, both of "regimes" and of "international relations," suggest such
an equivalence; see Young, "Regime Dynamics: The Rise and Fall of International Regimes," *International Organization* 36 (Spring 1982): 277–97. Also see his "International
Regimes: Problems of Concept Formation," *World Politics* 32 (April 1980): 331–56, and

litical game is to say that anything and everything goes. If this is all that we mean by "regimes," we have made no conceptual advance by using this term.[3]

Although I assume in this chapter that states are autonomous and independent, I derive an interest-based logic for the creation and maintenance of international regimes, which serve to circumscribe national behavior and so shape international interactions. Because it is theoretically rooted, the formulation can be used to delineate the nature and workings of regimes and to explain why and under what conditions they arise, how they are maintained and transformed, and when they may be expected to break down or dissolve. Further, it helps account for the fact that there are many different regimes rather than a single overarching one.[4]

Anarchy and Regimes

The conceptualization of regimes developed here is rooted in the classic characterization of international politics as relations between sovereign entities dedicated to their own self-preservation, ultimately able to depend only on themselves, and prepared to resort to force. Scholars often use anarchy as a metaphor to describe this state of affairs, providing an image of autonomous nation-states that consider every option

---

Compliance and Public Authority: A Theory with International Applications (Baltimore: Johns Hopkins University Press, 1979). My concern is to develop a conceptualization of regimes that delineates a subset of international politics.

3. Another extreme is when "regimes" are defined as "international institutions." In this sense, regimes equal the formal rules of behavior specified by the charters or constitutions of such institutions, and the study of regimes becomes no more than the study of international organizations.

4. The rest of this chapter originally appeared as "Coordination and Collaboration: Regimes in an Anarchic World," International Organization 36 (Spring 1982): 299–324. It is reprinted here with only slight changes, with the permission of MIT Press. I have, however, periodically cited articles that subsequently developed various arguments made in the original. For example, many of the points I made there were subsequently restated and expanded on by Duncan Snidal, "Coordination versus Prisoners' Dilemma: Implications for International Cooperation and Regimes," American Political Science Review 79 (December 1985): 923–42. In the original article, I argued that regimes were consistent with premises of autonomous self-interested behavior. Some saw my argument as inherently that of a liberal and labeled it, along with work by others, as "liberal institutionalism" or "neoliberalism." Others accepted the argument made in the original that I was beginning with realist premises and called it "modified structural realism" or "neorealism." But autonomous self-interested behavior underlies both classical realism and classical liberalism. What distinguishes liberals and realists is taken up in subsequent chapters of this book.

available to them and make their choices independently in order to maximize their own returns.

Any outcome that emerges from the interaction of states making independent decisions is a function of their interests and preferences. Depending on these interests, the outcome can range from pure conflict to no conflict at all and, depending on the actors' preference orderings, may or may not provide a stable equilibrium. Such independent behavior and the outcomes that result from it constitute the workings of normal international politics—not of regimes. An arms race, for example, is not a regime, even though each actor's decision is contingent on the other actor's immediately previous decision. As long as state behavior in the international arena results from unconstrained and independent decision making, there is no international regime.

A regime exists when the interaction between the parties is constrained or is based on joint decision making. Domestic society constitutes the most common regime. Even the freest and most open societies do not allow individualism and market forces full play; people are not free to choose from among every conceivable option—their choice set is constrained. The workings of a free market require a developed set of property rights, and economic competition is constrained to exclude predatory behavior.[5] Domestic society, characterized by the agreement of individuals to eschew the use of force in settling disputes, constitutes a regime precisely because it constrains the behavior of citizens.

Some argue that the advent of complex interdependence in the international arena means that states' actions are no longer unconstrained, that the use of force no longer remains a possible option. If the range of choice were indeed this circumscribed, we could in fact talk about the existence of an international regime similar to the domestic one. But if the international arena is one in which anything still goes, regimes will arise not because the actors' choices are circumscribed but because the actors eschew independent decision making.[6]

5. On the importance of property rights, see Thomas M. Carroll, David H. Ciscil, and Roger K. Chisholm, "The Market as a Commons: An Unconventional View of Property Rights," *Journal of Economic Issues* 13 (June 1979): 605–27. On the constrained sense of economic competition, see J. Hirshleifer, "Competition, Cooperation, and Conflict in Economics and Biology," *American Economic Review* 68 (May 1978): 238–43; Hirshleifer, "Economics from a Biological Viewpoint," *Journal of Law and Economics* 20 (April 1977): 1–52; and Hirshleifer, "Natural Economy versus Political Economy," *Journal of Social and Biological Structures* 1 (October 1978): 319–37.

6. The term "complex interdependence" is most fully presented in Robert O. Keohane

Figure 3. A no-conflict situation

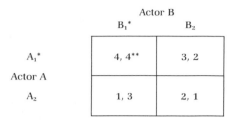

*Actor's dominant strategy
**Equilibrium outcome

International regimes exist when the patterned behavior of states results from joint rather than independent decision making.

International politics is typically characterized by independent self-
interested decision making, and states often have no reason to eschew
such individualistic behavior. There is no reason for a regime when
each state obtains its most preferred outcome by making independent
decisions, for there is simply no conflict. Examples include barter and
some forms of foreign aid (e.g., disaster relief). Figure 3 illustrates one
such situation, a case in which actors A and B both agree on a preferred
outcome, A_1B_1. In addition, both actors have a dominant strategy—a
course of action that maximizes an actor's returns no matter what the
other chooses. A prefers A_1 whether B chooses B_1 or B_2, and B prefers
B_1 regardless of A's decision. The result of their independent choices,
A_1B_1, is an equilibrium outcome, one from which neither actor can shift
unilaterally to better its own position.[7] The equilibrium outcome leaves
both actors satisfied. Because their interests are naturally *harmonious* and coincident, there is no conflict.[8] The actors reach what is for

and Joseph S. Nye, *Power and Interdependence: World Politics in Transition* (Boston: Little,
Brown, 1977); yet it remains unclear, for example, if the use of force remains an option
in the relations between advanced industrial societies but is dominated by other choices.
Alternatively, perhaps nations sometimes prefer to threaten the use of force on a contingent basis but recognize that the outcome resulting from the mutual use of force is the
least preferred outcome for all actors.

7. The A_1B_1 outcome is also a "coordination equilibrium," which David K. Lewis defines as an outcome from which neither actor can shift and make anyone better off; see
Convention: A Philosophical Study (Cambridge, Mass.: Harvard University Press, 1969),
p. 14.

8. This situation was subsequently dubbed harmonious interests by Robert O. Keohane, *After Hegemony: Cooperation and Discord in the World Political Economy* (Princeton, N.J.: Princeton University Press, 1984).

Figure 4. The assurance game

Actor B

	B_1	B_2
A_1	4, 4**	1, 3
A_2	3, 1	2, 2**

Actor A

**Equilibrium outcome

both the optimal result from their independent choices.[9] No regime is needed.

There is also no need for a regime when the actors share a most preferred outcome but neither has a dominant strategy. In Figure 4, A prefers A_1 only if B chooses B_1, and B prefers B_1 only if A chooses A_1. The equilibrium outcome that emerges, A_1B_1, leaves both satisfied. There is, however, a second equilibrium outcome possible in this case, one that emerges from each actor's desire to maximize its minimum gain. Such a minimax decision rule would leave A to choose A_2 and B to choose B_2, the course of action that would ensure that, at the very least, they avoid their worst outcomes. Yet the A_2B_2 outcome, although an equilibrium one, is mutually undesirable.[10] Thus as long as each actor is aware of the other's preferences, the two will converge on the A_1B_1 outcome that both most prefer. No regime is needed since both actors agree on a most preferred outcome, one that they can reach by acting autonomously.[11]

9. Individual accessibility is discussed by Jon Elster in *Ulysses and the Sirens: Studies in Rationality and Irrationality* (Cambridge: Cambridge University Press, 1979), p. 21.

10. Only A_1B_1, however, is a coordination equilibrium. The other equilibrium outcome, A_2B_2, does not qualify as such because each actor can shift from it and make the other better off by doing so. For Lewis, this does not pose a coordination problem, which requires the existence of two or more coordination equilibria; see *Convention*, p. 24.

11. For Elster, this case is individually inaccessible. Nonetheless, he expects convergence because the outcome is individually stable. I consider this case to be individually accessible precisely because there are convergent expectations. If regimes are understood to include any devices that help actors' expectations to converge, regimes might arise even in this case, although solely to provide information. The proffered information would provide each actor with assurance about the other's preferences, as would be necessary for expectations to converge on the one of the two equilibria that all prefer.

An emphasis on regimes as functional providers of information is found in Robert O. Keohane, "The Demand for International Regimes," *International Organization* 36 (Spring 1982): 325–55. I find Keohane's treatment of regimes and information problematic. He argues that given a demand for international agreements, the more costly the information,

Figure 5. An equilibrium outcome that leaves one actor aggrieved

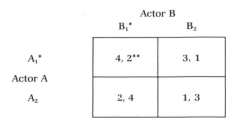

*Actor's dominant strategy
**Equilibrium outcome

The international extradition of criminals is an example of such an "assurance game." States began in the early nineteenth century unilaterally to adopt statutes stipulating extraditable offenses. Some states are satisfied with assurances of reciprocity before they agree to extradite criminal fugitives. Other states, however, are unsatisfied with such informal arrangements because of the political limitations that other nations may place on extradition. They require treaties to provide them with assurances that the other state will behave in a predictable fashion when questions of extradition arise.[12] It is important to understand, however, that these treaties provide only assurances and nothing more.

Nor will a regime arise when some actors obtain their most preferred outcome while others are left aggrieved. Figure 5 illustrates a situation in which both actors have dominant strategies leading to an equilibrium that is actor A's most preferred outcome but actor B's second-worst one. In such situations the satisfied actor has no reason to eschew independent decision making and the aggrieved actor would succeed only in making itself still worse off by being the only one to forgo

the greater the actual demand for international regimes (one of whose functions is to improve the information available to actors). This does not provide an adequate explanation for regimes as a form of international cooperation, for regimes presume a preexisting demand for agreements. It is this demand for agreements that requires explanation. Moreover, Keohane's formulation is too imprecise to specify what constitutes a regime. It is, for example, unclear whether he means to suggest that all mechanisms that provide information are examples of regimes even when the actors' interests are harmonious, or that they are not regimes because there is no demand for agreements in such cases. Since he presents the demand for agreements as a given, we do not know if the demand for information can be a basis for a demand for agreements or simply a basis for a demand for regimes, which assumes a demand for agreements.

12. In some cases, actors may require mechanisms for assurance, which extradition treaties exemplify. These treaties might thus be seen as "assurance regimes," regimes that arise when each actor's knowledge of others' preferences is enough to allow the actors' autonomous decisions to bring them to the outcome they all most prefer.

rational self-interested calculation. Voluntary export restraint is an example in which one actor gets its most preferred outcome while the other is left aggrieved by that equilibrium result.

In the foregoing examples, behavior and outcome result from the independent decisions of actors interacting in a context, prototypical of international relations, characterized by anarchy. There are situations, however, in which all the actors have an incentive to eschew independent decision making: situations, that is, in which individualistic self-interested calculation leads them to prefer joint decision making because independent self-interested behavior can result in undesirable or suboptimal outcomes. I refer to these situations as dilemmas of common interests and dilemmas of common aversions.[13]

Dilemmas of Common Interests

The dilemma of common interests arises when independent decision making leads to equilibrium outcomes that are Pareto-deficient—outcomes in which all actors prefer another given outcome to the equilibrium outcome. The classic example is, of course, the prisoners' dilemma. Figure 6 illustrates the two-actor prisoners' dilemma in which both actors prefer the A_1B_1 outcome to the A_2B_2 equilibrium. But the preferred A_1B_1 outcome is neither individually accessible nor stable. To arrive at the Pareto-optimal outcome requires that all actors eschew their dominant strategy. In addition, they must not greedily attempt to obtain their most preferred outcome once they have settled at the unstable outcome they prefer to the stable equilibrium.[14]

The prisoners' dilemma is used as an allegory for a variety of situations. It is, for instance, the classic illustration of the failure of market forces always to result in optimal solutions—that is, of market rationality leading to suboptimal outcomes. Oligopolists, for example, prefer collusion to the deficient equilibrium that results from their competition.[15] Ironically, government intervenes to prevent collusion and en-

13. The conceptualization of regimes presented here, that they arise to deal with dilemmas of common interests and common aversions, is not therefore based on any inherent notion of "principles." Indeed, it is easy to conceive of unprincipled regimes, such as OPEC. Regimes may, but need not, have some principle underlying them.

14. The prisoners' dilemma is the only two-actor example of a Pareto-deficient equilibrium that occurs when both actors have dominant strategies. It is for this reason that it has received so much attention from scholars.

15. The role of game models in analyzing oligopolistic relations is described by Jesse W. Markham, "Oligopoly," in *International Encyclopedia of the Social Sciences*, vol. 11

Figure 6. Prisoners' dilemma

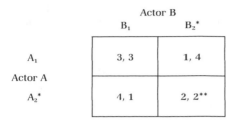

*Actor's dominant strategy
**Equilibrium outcome

force the outcome that is suboptimal for the oligopolists. There are other situations of suboptimality, such as problems of collective goods and externalities, that lead to government intervention, but in this case to ensure collusion and collaboration and thus to ensure avoidance of the suboptimal equilibrium outcome.[16]

Political theorists use the prisoners' dilemma to explain the contractarian-coercion conjunction at the root of the modern state, arguing that the state of nature is a prisoners' dilemma in which individuals have a dominant strategy of defecting from common action but in which the result of this mutual defection is deficient for all. Yet the outcome that results from mutual cooperation is not an equilibrium one since each actor can make itself immediately better off by cheating. It is for this reason, political theorists argue, that individuals came together to form the state by agreeing to coerce one another and thus ensure the optimal outcome of mutual cooperation. In other words, they agreed to coerce one another in order to guarantee that no individual would take advantage of another's cooperation by defecting from the pact and refusing to cooperate. States are thus coercive institutions that allow individuals to eschew their dominant strategies—an individual actor's rational course—as a matter of self-interest in order to ensure an optimal rather than a Pareto-deficient equilibrium outcome.[17]

(New York: Macmillan, 1968), pp. 283–88. F. M. Scherer discusses the prisoners' dilemma as a model for oligopolistic interaction in *Industrial Market Structure and Economic Performance* (Chicago: Rand McNally, 1970). The same observation is made by Lester G. Telser, who redubs the prisoners' dilemma as it applies to oligopolies the "cartel's dilemma"; see *Competition, Collusion, and Game Theory* (Chicago: Aldine-Atherton, 1972), p. 143.

16. For a general discussion of suboptimality, see Jon Elster, *Logic and Society: Contradictions and Possible Worlds* (New York: John Wiley, 1978), pp. 122–34.

17. In one formulation, Jon Elster defines "politics" as "the study of ways of transcending the Prisoners' Dilemma"; see "Some Conceptual Problems in Political Theory,"

Put more simply, the argument is that individuals come together to form the state in order to solve the dilemma of common interests. The existence of a prisoners' dilemma preference ordering creates the likelihood that individual rationality will lead to suboptimal outcomes, a classic case of market failure. Individuals have a common interest in constraining the free reign of their individuality and independent rationality, and they form domestic political regimes to deal with the problem.

This view of the state is reinforced by the literature on collective goods, in which scholars argue that the suboptimal provision of collective goods stems from the individual's incentive to be a free rider, to enjoy the benefits of goods characterized by nonexcludability. Under certain conditions, the problem of collective goods is a classic prisoners' dilemma in which each individual is better off not contributing to the provision of a collective good, but in which the equilibrium outcome of everyone's deciding to be a free rider is a world in which all are worse off than if they had contributed equally to the provision of the good.[18] Some in fact argue that the state is formed to ensure the provision of collective goods; the state coerces contributions from all individuals, each of whom would rather be a free rider but goes along because of the guarantee that all others will be similarly coerced. They form the state because the alternative outcome is a Pareto-deficient world in which collective goods are not provided. The most basic collective good provided by the state is, of course, security from outside attack. Thus we have an explanation for the rise of states that also illuminates the anarchic character of relations between these states. The anarchy that engenders state formation is tamed only within domestic society. Individuals sacrifice a certain degree of autonomy—but the newly established nations do not do so. A world of vying individuals is replaced by a world of vying nations.

Regimes in the international arena are also created to deal with the

in *Power and Political Theory: Some European Perspectives*, ed. Brian Barry (London: John Wiley, 1976), pp. 248–49. Laurence S. Moss provides an assessment of modern and somewhat formal equivalents to the Hobbesian and Lockean views of state formation in "Optimal Jurisdictions and the Economic Theory of the State: Or, Anarchy and One-world Government Are Only Corner Solutions," *Public Choice* 35 (1980): 17–26. See also Michael Taylor, *Anarchy and Cooperation* (London: John Wiley, 1976). Elster criticizes Taylor's alternative in *Logic and Society*, pp. 156–57, and *Ulysses and the Sirens*, pp. 64, 143, 146.

18. Russell Hardin, "Collective Action as an Agreeable n-Prisoners' Dilemma," *Behavioral Science* 16 (September 1971): 472–81. Note Elster's distinction between counterfinality and suboptimality in explaining the behavior of free riders; *Logic and Society*, pp. 122–23.

collective suboptimality that can emerge from individual behavior.[19] There are, for example, international collective goods whose optimal provision can be ensured only if states eschew the independent decision making that would otherwise lead them to be free riders and would ultimately result in either the suboptimal provision or the nonprovision of the collective good. One such problem of international politics, that of collective security, was in fact the focus of some of the earliest studies of collective goods.[20]

Collective goods issues are not the only problems characterized by prisoners' dilemma preferences for which international regimes can provide a solution.[21] The attempt to create an international trade regime after World War II was, for example, a reaction to the results of the beggar-thy-neighbor policies of the depression years. All nations would be wealthier in a world that allows goods to move unfettered across national borders. Yet any single nation, or group of nations, could improve its position by cheating—erecting trade barriers and restricting imports.[22] The state's position remains improved only as long as other

19. The dilemmas discussed in this chapter refer to specific actors and not necessarily to the system as a whole. In the prisoners' dilemma, for example, only the prisoners themselves face a Pareto-deficient outcome. The rest of society finds the outcome of their dilemma to be optimal. This is precisely analogous to the situation of oligopolists, who prefer collusion to competition. The rest of society, however, would prefer that they compete rather than collude. The collective suboptimality need not necessarily exist for all actors in the system.

20. This literature was spawned by Mancur Olson, Jr., and Richard Zeckhauser, "An Economic Theory of Alliances," *Review of Economics and Statistics* 48 (August 1966): 266–79. Other essays linking collective goods and international cooperation include Bruce M. Russett and John D. Sullivan, "Collective Goods and International Organization," *International Organization* 25 (Autumn 1971): 845–65; John Gerard Ruggie, "Collective Goods and Future International Collaboration," *American Political Science Review* 66 (September 1972): 874–93; and Todd Sandler and Jon Cauley, "The Design of Supranational Structures: An Economic Perspective," *International Studies Quarterly* 21 (June 1977): 251–76. More recent work stresses different institutional arrangements for international collective goods. See Todd M. Sandler, William Loehr, and Jon T. Cauley, "The Political Economy of Public Goods and International Cooperation," *Monograph Series in World Affairs* 15 (1978), and Duncan Snidal, "Public Goods, Property Rights, and Political Organizations," *International Studies Quarterly* 23 (December 1979): 532–66.

21. That is, although collective goods problems may be prisoners' dilemmas, not all prisoners' dilemmas are problems of collective goods. Some even argue that not all collective goods problems are prisoners' dilemmas, see, among others, Irwin Lipnowski and Shlomo Maital, "Voluntary Provision of a Pure Public Good as the Game of 'Chicken,'" *Journal of Public Economics* 20 (April 1983): 381–86; and Michael Taylor and Hugh Ward, "Chickens, Whales, and Lumpy Goods: Alternate Models of Public-Goods Provision," *Political Studies* 30 (September 1982): 350–70. Also see the new book by George Tsebelis, *Nested Games: Rational Choice in Comparative Politics* (Berkeley and Los Angeles: University of California Press, 1990).

22. Indeed, international trade regimes have historically exemplified the subsystemic character of many regimes. Scholars often characterize the middle of the nineteenth

nations do not respond in kind. Such a response is, however, the natural course for those other nations. When all nations pursue their dominant strategies and erect trade barriers, however, they can engender the collapse of international trade and depress all national incomes.[23] That is what happened in the 1930s, and what nations wanted to avoid after World War II.[24]

Dilemmas of Common Aversions

Regimes also provide solutions to dilemmas of common aversions. Unlike dilemmas of common interests, in which the actors have a common interest in ensuring a particular outcome, dilemmas of common aversions are characterized by actors' having a common interest in avoiding a particular outcome. These situations occur when actors with contingent strategies do not most prefer the same outcome but do agree that there is at least one outcome that all want to avoid. These criteria define a set of situations with multiple equilibria (two equilibria if there are only two actors, each with two choices) in which coordination is required if the actors are to avoid that least preferred outcome.[25] Thus

century, for example, as the era of free trade. Yet several major states, including the United States and Russia, did not take part. Similarly, the post-1945 era is now commonly referred to as the period of American economic hegemony. Ironically, this characterization is of a postwar economic system established by and within the sphere of only one pole of a bipolar international system—a bipolarity that has typically been offered as the most important characterization of the age. In other words, we should continually be reminded that references to "the" postwar economic system are, in fact, to a subsystem that excludes the Soviet bloc. See Arthur A. Stein, "The Hegemon's Dilemma: Great Britain, the United States, and the International Economic Order," *International Organization* 38 (Spring 1984): 355–86.

23. The discussion here is quite carefully worded, because trade policy is not a collective good. States can discriminate and exclude specific countries from the benefits of trade liberalization. This point is discussed in Stein, "The Hegemon's Dilemma."

24. A similar argument can sometimes be made about the decision to devalue a currency or maintain par value in a fixed-exchange-rate system when devaluation, which is every nation's dominant strategy, nevertheless results in the suboptimal outcome of mutual devaluation. Richard N. Cooper uses simple games in his discussion of the choice of an international monetary regime; see "Prolegomena to the Choice of an International Monetary System," *International Organization* 29 (Winter 1975): 63–97.

25. In the dilemma of common interests, actors are averse to the suboptimal equilibrium outcome, and resolution involves their arriving at the outcome they prefer to the equilibrium one. The dilemma is their inability individually to arrive at the outcome they prefer to the equilibrium one. In the dilemma of common aversions, on the other hand, the actors do have a common interest in avoiding a particular outcome, but their dilemma is the possibility that they might arrive at a mutual aversion without some coordination. Beyond their desire to avoid that aversion, however, they disagree about which of the multiple equilibria they prefer.

Figure 7. Dilemma of common aversions and common indifference

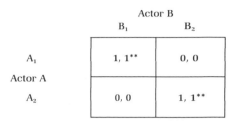

In this example, 1 = most preferred, 0 = least preferred
**Equilibrium outcome

such dilemmas can also lead to the formation of regimes by providing the incentive for nations to eschew independent decision making.

Figure 7 provides one example of a dilemma of common aversions. Neither actor in this situation has a dominant strategy; nor does either most prefer a single given outcome. Rather, there exist two outcomes that both actors value equally and two outcomes that both wish to avoid. Thus the situation has two equilibria, A_1B_1 and A_2B_2, but since the actors have contingent strategies, they cannot be certain that they will arrive at one of these outcomes if they act independently and simultaneously. Without coordination, they may well end up with one of the outcomes that neither wants.[26]

This example of common aversions is relatively easy to deal with because the actors do not have divergent interests; neither cares which of the two equilibria emerges. Any procedure that allows for a convergence of their expectations makes coordination possible by allowing the actors to arrive at one of the equilibria. It is in such situations that conventions play an important role. Driving on the right is a simple coordination mechanism that allows for the smooth movement of traffic in opposite directions without collisions and bottlenecks. It is an arbitrary convention that allows actors' expectations to converge on one of the equilibrium outcomes. The alternative convention of driving on the left permits coordination by convergence on the other equilibrium. The actors are indifferent between the two equilibria.

There are times, however, when, although both still agree on the least preferable outcome or outcomes, each prefers a different one of the possible equilibria. In Figure 8, for example, there are two equilibria $(A_2B_1$ and $A_1B_2)$, both of which the actors prefer over either of the other

26. Both equilibria are also coordination equilibria. In this case there is no minimax solution.

Figure 8. Dilemma of common aversions and divergent interests

Actor B

	B₁	B₂
A₁	2, 2	3, 4**
A₂	4, 3**	1, 1

Actor A

**Equilibrium outcome

possible outcomes. Each does, however, most prefer one of the two equilibria—although they do not most prefer the same one. Actor A prefers A_2B_1, whereas B favors A_1B_2.[27]

When actors confront mutual aversions but diverge in their assessments of the equilibria, coordination can be accomplished in two different ways. In either case the coordination regime establishes rules of behavior that allow actors' expectations to converge whenever the dilemma arises. One means of ensuring coordination is to specify behavior according to actors' characteristics. Alternatively, the prearrangement can specify behavior by context. One example of this dilemma is provided by the simultaneous arrival of a north- or southbound and an east- or westbound car at an intersection. In this case, both drivers most want to avoid a collision. They would also prefer not to sit at their corners staring at each other. There are two ways for them to sit at the intersection safely: either A goes first, or B does. The problem is that neither wants to be the one to wait. A coordination rule based on actors' characteristics would specify, for example, that Cadillacs drive on while Volkswagens sit and wait. Under such a regime, more likely than not "coordination for the powerful," the same actor always gets the equilibrium that it prefers. Alternatively, the actors could adopt a contextual rule; one example is the specification that the actor on the right always gets the right of way. In this case the context determines whether any actor gets its more preferred equilibrium; sometimes it does, and sometimes not. Ideally, this "fairness doctrine" would ensure that all actors get their most preferred equilibrium half the time.

27. If each of the actors chooses its minimax option, the A_1B_1 outcome results. This outcome is not their mutual aversion, but it is a Pareto-deficient nonequilibrium outcome because both prefer it less than either equilibrium.

Collaboration and Coordination

Regimes arise because actors forgo independent decision making in order to deal with the dilemmas of common interests and common aversions. They do so in their own self-interest, for in both cases jointly accessible outcomes are preferable to those that are or might be reached independently. It is in their interests mutually to establish arrangements to shape their subsequent behavior and allow expectations to converge, thus solving the dilemmas of independent decision making.[28] Yet the need to solve the dilemmas of common interests and aversions provides two different bases for international regimes, which helps explain certain differences between regimes that have often confused analysts. Regimes established to deal with the dilemma of common interests differ from those created to solve the dilemma of common aversions. The former require collaboration, the latter coordination.

The dilemma of common interests occurs when there is only one equilibrium outcome that is deficient for the involved actors. In other words, this dilemma arises when the Pareto-optimal outcome that the actors mutually desire is not an equilibrium outcome. To solve such dilemmas and guarantee the Pareto-optimal outcome, the parties must collaborate, and all regimes intended to deal with dilemmas of common interest must specify strict patterns of behavior and ensure that no one cheats.[29] Because each actor requires assurances that the other will

28. Precommitment has been variously described as the power to bind, as imperfect rationality, and as egonomics; see Thomas C. Schelling, *The Strategy of Conflict* (Cambridge, Mass.: Harvard University Press, 1960), pp. 22–28; Elster, *Ulysses and the Sirens*, pp. 36–111; and T. C. Schelling, "Egonomics, or the Art of Self Management," *American Economic Review* 68 (May 1978): 290–94. Such a formulation of prior agreement on principles does not require John Rawls's veil of ignorance; see *A Theory of Justice* (Cambridge, Mass.: Harvard University Press, 1971). Thinking ahead without agreement in strategic interaction, however, is no solution; see Frederic Schick, "Some Notes on Thinking Ahead," *Social Research* 44 (Winter 1977): 786–800.

29. The prisoners' dilemma is the only situation with a Pareto-deficient equilibrium in which all the actors have dominant strategies. There are other cases of Pareto-deficient equilibria in which some have dominant strategies and some contingent strategies. These, too, are dilemmas of common interests and require regimes for solution; in these cases, however, only those actors with dominant strategies must eschew independent decision making. Thus the regime formed to ensure collaboration in this case is likely to have stipulations and requirements that apply asymmetrically to those who must eschew independent decision making to achieve optimality and to those who must be assured that others have actually done so and will continue to do so.

Some argue that the cooperative nonequilibrium outcome of the prisoners' dilemma can emerge spontaneously—without collaborative agreement. Social psychologists have done extensive experiments on the emergence of cooperation in repeated plays of the

40					WHY NATIONS COOPERATE

also eschew its rational choice, such collaboration requires a degree of formalization. The regime must specify what constitutes cooperation and what constitutes cheating, and each actor must be assured of its own ability to spot immediately others' cheating.

The various SALT agreements provide examples of the institutionalized collaboration required in a regime intended to deal with the dilemma of common interests, for the security dilemma is an example of a prisoners' dilemma in which all actors arm themselves even though they prefer mutual disarmament to mutual armament. Yet international disarmament agreements are notoriously problematic. Indeed, the decision to comply with or cheat on an arms control agreement is also a prisoners' dilemma in which each actor's dominant strategy is to cheat. Thus it is not surprising that arms control agreements are highly institutionalized, for these regimes are continually concerned with compliance and policing.[30] They must define "cheating" quite explicitly, ensure that it be observable, and specify verification and monitoring procedures.

Oligopolists also confront the dilemma of common interests, and their collusion represents the collaboration necessary for them to move from the suboptimal equilibrium that would otherwise result. Such collusive arrangements require policing and monitoring because of the individual's incentive to cheat. International market-sharing arrange-

prisoners' dilemma game; for reviews see Dean G. Pruitt and Melvin J. Kimmel, "Twenty Years of Experimental Gaming: Critique, Synthesis, and Suggestions for the Future," *Annual Review of Psychology* 28 (1977): 363–92; and David M. Messick and Marilynn B. Brewer, "Solving Social Dilemmas: A Review," in *Review of Personality and Social Psychology*, vol. 4, ed. Ladd Wheeler and Phillip Shaver (Beverly Hills, Calif.: Sage Publications, 1983), pp. 11–44. See also Anatol Rapoport, Melvin J. Guyer, and David G. Gordon, *The 2 × 2 Game* (Ann Arbor: University of Michigan Press, 1976). For a mathematician's deductive assessment of the prospects for the emergence of such cooperation, see Steve Smale, "The Prisoner's Dilemma and Dynamical Systems Associated to Non-Cooperative Games," *Econometrica* 48 (November 1980): 1617–34. See also Robert Axelrod, "The Emergence of Cooperation among Egoists," *American Political Science Review* 75 (June 1981): 306–18.

The conditions for this cooperation are rarely met in international politics, however. The first such requirement is that play be repeated indefinitely. Because states can disappear, and because they are therefore concerned with their own survival, international politics must be seen by the actors as a finite game (see Chapter 4). Moreover, the stakes in international politics are typically so high that fear of exploitation will ensure that states follow their dominant strategy, to defect, in the absence of a collaborative agreement. The importance of the relative desirability of cheating for the prospects for collaboration is discussed in Robert Jervis, "Cooperation under the Security Dilemma," *World Politics* 30 (January 1978): 167–214. These observations also became the basis for an article by Charles Lipson, "International Cooperation in Economic and Security Affairs," *World Politics* 37 (October 1984): 1–23.

30. This standard view of arms issues as prisoners' dilemmas is explicitly challenged in Chapter 5.

ments exemplify this collusive form of collaboration and require the same sort of monitoring provisions. Not surprisingly, such successful market-sharing regimes as the International Coffee Agreement have extensive enforcement provisions and elaborate institutional structures for monitoring compliance.[31]

"The tragedy of the commons" exemplifies the dilemma of common interests. The commons were pasture, grazing grounds open to all, and the tragedy was that overgrazing resulted from unrestrained individual use. This is not, as it may seem at first, a dilemma of common aversions in which the actors' least preferred outcome is the depletion of a valuable common resource. Rather, each actor most prefers to be the only user of a common resource, next prefers joint restraint in the mutual use of the good, then prefers joint unrestrained use even if it leads to depletion, and least prefers a situation in which its own restraint is met by the other actors' lack of restraint. Each actor would rather share in such use of the resource that leads to depletion than to see its own restraint allow either the continued existence of the resource for others' use or the disappearance of the resource because the others show no restraint. The actors have a common interest in moving from their suboptimal (but not least preferred) outcome to one in which they exercise mutual restraint by collaboratively managing the resource. The commons thus represent a class of dilemmas of common interests in which individually rational behavior leads to a collectively suboptimal outcome.[32] Current international commons problems, such as the overfishing of a common sea, are all international manifestations of this dilemma of common interests.

By contrast, regimes intended to deal with the dilemma of common aversions need only facilitate coordination. Such situations have multiple equilibria, and these regimes do not have to guarantee a particular outcome or compliance with any specific course of action, for they are created only to ensure that particular outcomes be avoided.[33] Never-

31. Bart S. Fisher, *The International Coffee Agreement: A Study in Coffee Diplomacy* (New York: Praeger, 1972), and Richard B. Bilder, "The International Coffee Agreement: A Case History in Negotiation," *Law and Contemporary Problems* 28 (Spring 1963): 328–91. The latter appeared in a special issue devoted to "International Commodity Agreements."

32. Garrett Hardin, "The Tragedy of the Commons," *Science*, December 13, 1968, pp. 1243–48; Thomas C. Schelling characterizes the commons as a prisoners' dilemma in *Micromotives and Macrobehavior* (New York: Norton, 1978), pp. 110–15.

33. The following authors all discuss coordination, although they do not agree fully on a definition: Schelling, *Strategy of Conflict*; Lewis, *Convention*; Philip B. Heymann, "The Problem of Coordination: Bargaining and Rules," *Harvard Law Review* 86 (March 1973): 797–877; and Robert E. Goodin, *The Politics of Rational Man* (London: John Wiley, 1976), pp. 26–46. Also see Andrew Schotter, *The Economic Theory of Social Institutions* (Cam-

theless, such coordination is difficult to achieve when both actors disagree in the choice of preferred equilibrium. The greater this conflict of interest, the harder it is for them to coordinate their actions. Yet once established, the regime that makes expectations converge and allows the actors to coordinate their actions is self-enforcing; any actor that departs from it hurts only itself.[34] Thus there is no problem here of policing and compliance. Defections do not represent cheating for immediate self-aggrandizement but are expressions of relative dissatisfaction with the coordination outcome. An actor will threaten to defect without actually doing so; it may choose to go through with its threat only if the other actor does not accede to its demands. Again, such defection is never surreptitious cheating; it is a public attempt, made at some cost, to force the other actor into a different equilibrium outcome. Departures from regime-specified behavior thus represent a fundamentally different problem in coordination regimes than in collaboration ones.

There are many international regimes that serve to facilitate coordination and thus solve the dilemma of common aversions. These solutions provide mechanisms that allow actors' expectations to converge on one of the possible equilibria. Conventions alone are adequate in these situations; institutions are not required. Not surprisingly, many involve standardization.[35] The adoption of a common gauge for railroad tracks throughout Western Europe is one example.[36]

---

bridge: Cambridge University Press, 1981); Russell Hardin, *Collective Action* (Baltimore: Johns Hopkins University Press, 1982); and Robert Sugden, *The Economics of Rights, Cooperation, and Welfare* (New York: Basil Blackwell, 1986). The distinction between collaboration and coordination made here can be compared to distinctions between negative and positive *coordination* and between negative and positive *cooperation* made by the following: Marina v. N. Whitman, "Coordination and Management of the International Economy: A Search for Organizing Principles," in *Contemporary Economic Problems 1977*, ed. William Fellner (Washington, D.C.: American Enterprise Institute for Public Policy Research, 1977), p. 321; and Jacques Pelkmans, "Economic Cooperation among Western Countries," in *Challenges to Interdependent Economies: The Industrial West in the Coming Decade*, ed. Robert J. Gordon and Jacques Pelkmans (New York: McGraw-Hill, 1979), pp. 97–123.

34. This notion of self-enforcement differs from that developed by L. G. Telser, "A Theory of Self-enforcing Agreements," *Journal of Business* 53 (January 1980): 27–44. For Telser, an arrangement is self-enforcing if the actor calculates that defection may bring future costs. Thus even if cheating brings immediate rewards, an actor will not cheat if others' responses cause it to bear a net loss. For me, regimes are self-enforcing only if the cost that an actor bears for defecting is immediate rather than potential and is brought about by its own defection rather than by the response of others to that defection.

35. See, for example, Charles P. Kindleberger, "Standards as Public, Collective, and Private Goods," *Kyklos* 36 (1983): 377–96.

36. Standardization may reflect harmonious interests rather than coordination solu-

Traffic conventions are also examples of international regimes.[37] Under the rules of the International Civil Aviation Organization, for example, every flight control center must always have enough English-speakers on duty to direct all those pilots who do not speak the native language of the country whose airspace they happen to be crossing.[38] Communication between ground and aircraft may be in any mutually convenient language, but there must be a guarantee that communication is indeed possible; finding a language matchup cannot be left to chance. Thus English is recognized as the international language of air traffic control, and all pilots who fly between nations must speak enough English to talk to someone in the flight control center. The pilot who never leaves French airspace is perfectly safe knowing only French, and should a Mexicana Airlines pilot wish to speak Spanish to the air traffic controller in Madrid, that is also acceptable. But if no air traffic controller speaks the pilot's language, the parties can always converse in English. The mutual aversion, an air disaster, is avoided, and a safe equilibrium is ensured.[39]

Preemption provides still another solution to dilemmas of common aversions. In these situations with multiple equilibria and a mutually least preferred outcome, an actor's incentive is often to preempt the other because it knows that the other must then defer to it. If it is wrong, however, if an oncoming car fails to swerve while it also keeps going, the attempted preemption leads directly to the common aversion. Often, however, preemption is based on firm knowledge or safe assumptions and is therefore successful. In these cases preemption

tions to dilemmas of common aversions. This may, for example, explain the adoption of a common calendar.

37. Schelling provides an interesting discussion of the traffic light as a self-enforcing convention in *Micromotives and Macrobehavior*, pp. 119–21.

38. The organization is the governing body for almost all international civil air traffic.

39. There does exist a dilemma of common aversions that can be solved either by coordination or by collaboration. As in other situations characterized as dilemmas of common aversions, the actors in the game of chicken have contingent strategies, do not agree on a most preferred outcome, but do share a mutual aversion. In this case, the actors diverge in their assessment of the two equilibria. Unlike those of other dilemmas of common aversions, the two equilibria in chicken are not coordination equilibria. In chicken, the nonequilibrium minimax outcome is the second choice of both actors and is not Pareto-deficient. Thus the situation is not merely one of deadlock avoidance but one that can be solved either by coordination to arrive at one of the two equilibria or by collaboration to accept second-best. Here, too, the collaboration is not self-enforcing and requires mutual assurances about defection from a particular outcome. No-fault insurance agreements are one example of a collaboration regime to resolve a dilemma of common aversions. Note that Lewis would not consider chicken to be a coordination problem because the two equilibria in chicken are not coordination equilibria. I believe that it is a coordination problem, but one that collaboration can also solve.

forms the basis of coordination, and it works well when it involves the exercise of squatters' rights in an area where they are traditionally respected or are likely to be so. One striking example has been the preemption of the radio frequencies within accepted constraints. International meetings have allocated various portions of the radio spectrum for specific uses, and countries have then been free to broadcast appropriately along whatever frequency is available. They are required to register the frequencies they have claimed with the International Frequency Registration Board. Nevertheless, other nations sometimes broadcast on those same wavelengths when they are available, a practice that is not permitted but is accepted. It has been without challenge that the Soviet Union, for example, prowls the shortwave band for unused frequencies on which it then broadcasts its own propaganda. The result is a system of allocation that allows all nations the use of an adequate number of frequencies for broadcast with minimal international interference.

As the number of nations in the world has increased, however, the radio band has become more crowded, and third-world nations have demanded greater access to radio frequencies.[40] To some, the allocation of frequencies has now become a dilemma of common interests, for their worst outcome is to fail to get on the radio at all. In other words, they actually prefer the radio traffic jam that previously constituted a dilemma of common aversions in the hope that the other broadcaster will eventually give up and leave them an unimpeded signal. No longer willing to accept what has become in practice a form of coordination for the powerful, they are calling for "planning" (i.e., collaboration) to replace the current system.

Regimes and Interests

This conceptualization of regimes is interest-based. It suggests that the same forces of autonomously calculated self-interest that lie at the root of the anarchic international system also provide the foundation for international regimes as a form of international order. The same forces that lead individuals to bind themselves together to escape the state of

40. For background and analysis of the World Administrative Radio Conference of 1979, see the articles in *Foreign Policy*, no. 34 (Spring 1979): 139–64, and those in *Journal of Communication* 29 (Winter 1979): 143–207. See also "Scramble for the Waves," *Economist*, September 1, 1979, p. 37; "The Struggle over the World's Radio Waves Will Continue," *Economist*, December 8, 1979, p. 83; and "Policing the Radio," *New Statesman*, December 14, 1979, p. 924.

nature also lead states to coordinate their actions, even to collaborate with one another. Quite simply, there are times when rational self-interested calculation leads actors to abandon independent decision making in favor of joint decision making.[41]

This formulation presumes the existence of interdependence—that an actor's returns are a function of others' choices as well as its own. If actors were independent in the sense that their choices affected only their own returns and not others', there would be no basis for international regimes.[42] Interdependence in the international arena, especially given the relatively small size of the system, makes mutual expectations (and therefore perceptions) very important.[43] An analogy from economics is often used to make this point. There are so many firms in a perfectly competitive market that each firm is assumed to have a dominant strategy and to make decisions without taking into account expectations of others' potential behavior or responses. Oligopolistic or imperfect competition is distinguished precisely by the small number of actors, which makes necessary and possible the incorporation of expectations in the context of interdependence.

This conceptualization also explains why the same behavior that sometimes results from independent decision making can also occur under regimes. Arms buildups provide one example. On the one hand, an arms race is not a regime, despite the existence of interaction and although each actor's decisions are contingent on the other's. An arms race is not a regime because the behavior, although patterned, is the result of independent decision making. On the other hand, arms increases can result from an arms control agreement that is a regime because the arms buildup results from mutual arrangements that shape subsequent decisions. Indeed, most arms control agreements have been not arms reductions agreements but agreements of controlled escalation. By arriving at such an agreement, both actors thus participate in shaping their subsequent actions.[44]

41. For a philosophical treatment that characterizes similar choices as constrained maximization and as the basis for morals by agreement, see David Gauthier, "Reason and Maximization," *Canadian Journal of Philosophy* 4 (March 1975): 411–33, and his *Morals by Agreement* (New York: Oxford University Press, 1986).

42. The absence of regimes does not mean, however, that the actors are independent of one another.

43. The conditions in which misperception matters, and the ways in which it matters, are delineated in Chapter 3.

44. Goodin, in *Politics of Rational Man*, p. 26, puts it this way: "Joint decision making is said to occur when all actors participate in determining the decisions of each actor. It implies that there was interaction between all the actors prior to the decisions and that this interaction shaped the decision of each actor." It is not surprising, then, that

This conceptualization of regimes also clarifies the role of international institutions, which many equate with regimes. Even those who recognize that regimes need not be institutionalized still suggest that institutionalization is one of their major dimensions. In fact, one scholar refers to noninstitutionalized regimes as quasi regimes.[45] But the conceptualization I have presented here suggests that international organizations and regimes are independent of one another; each can exist without the other. Regimes can be noninstitutionalized as well as institutionalized, and international organizations need not be regimes, although they certainly can be.[46] The United Nations is an example of an international organization that is not a regime, for mere membership in no way constrains independent decision making. The UN provides a forum for formal and informal interaction and discussion, but it is not a regime because membership generates no convergent expectations that constrain and shape subsequent actions.

The presumption of the existence of the dilemmas of common interests and common aversions that give rise to regimes assumes that self-interested actors do indeed have things in common. This is very much a liberal, not mercantilist, view of self-interest; it suggests that actors focus on their own returns and compare different outcomes with an eye to maximizing their own gains.

An alternative conception of competitive self-interest is that actors seek to maximize the difference between their own returns and those of others. This decision rule, that of difference maximization, is competitive, whereas a decision criterion of self-maximization is individualistic. When applied by any actor, it transforms a situation into one of pure conflict in which the actors have no mutual interests or common

two recent formulations both stress the importance of agreement as part of their definition of "regimes": see Young, "International Regimes"; and Ernst B. Haas, "Why Collaborate? Issue Linkage and International Regimes," *World Politics* 32 (April 1980): 358. For interesting delineations of the range of decision making procedures, see Knut Midgaard, "Co-operative Negotiations and Bargaining: Some Notes on Power and Powerlessness," in *Power and Political Theory*, ed. Barry, pp. 117–37; and I. William Zartman, "Negotiations as a Joint Decision-Making Process," *Journal of Conflict Resolution* 21 (December 1977): 620–23. Both of these authors, however, emphasize the bargaining process. Various forms of international cooperation can also be seen as forms of decision making; see Jan Tinbergen, "Alternative Forms of International Co-operation: Comparing Their Efficiency," *International Social Science Journal* 30 (1978): 224–25.

45. Hayward R. Alker, Jr., "A Methodology for Design Research on Interdependence Alternatives," *International Organization* 31 (Winter 1977): 37–38.

46. Although I do not define regimes by reference to their degree of institutionalization, it is true that collaboration regimes are more likely to be institutionalized than coordination regimes, because of the requirements of enforcement.

aversions; it implies a constant-sum world in which an improvement in one actor's returns can come only at the expense of another's.[47]

Actors that are competitors rather than individualists do not confront dilemmas of common interests or common aversions. Out for relative gain, they have nothing in "common." The prisoners' dilemma is an interesting illustration of this point. When both actors apply a difference maximization decision rule to the preference ordering that defines a prisoners' dilemma, the situation that results is one in which the actors' dominant strategies are the same. They no longer find the equilibrium outcome deficient and do not prefer an alternative one. The situation no longer provides them with a rational incentive to eschew independent decision making in order to create and maintain a regime. Thus, to see the existence of international regimes composed of sovereign entities that voluntarily eschew independent decision making in certain cases is to see the world in nonconstant-sum terms, a world in which actors can have common interests and common aversions.[48] It is self-interested actors that find a common interest in eschewing individuality to form international regimes.

This conceptualization of regimes also explains why there are so many regimes and why they vary in character, why they exist in some issue areas and not in others, and why states will form regimes with one another in one domain while they are in conflict in another. The existence or nonexistence of regimes to deal with given issues, indeed the very need to distinguish them by issue, can be attributed to the existence of different constellations of interests in different contexts.

Structural Bases of Regime Formation

In this formulation, the factors that others argue to be the bases of regime formation, whatever they may be, should be understood instead as constituting the determinants of those different patterns of interests

47. Difference maximization is discussed in Chapter 5 and by Charles G. McClintock, "Game Behavior and Social Motivation in Interpersonal Settings," in *Experimental Social Psychology*, ed. Charles Graham McClintock (New York: Holt Rinehart and Winston, 1972), pp. 271–92. Taylor calls them pure difference games and designates them a subtype of games of difference generally; see *Anarchy and Cooperation*, pp. 73–74. See also Martin Shubik, "Games of Status," *Behavioral Science* 16 (March 1971): 117–29.

48. Those who argue that world politics constitutes a zero-sum game cannot, of course, sustain their position at the extremes. After all, it is impossible for all dyadic relationships to be zero-sum or constant-sum in a world of more than two actors. Thus, even if some relationships in international politics are zero- or constant-sum, there must also exist some subset of relationships that are nonconstant-sum and which hence provide a basis for regime formation among this subset of nations. See Chapters 5 and 6.

that underlie the regimes themselves. More specifically, I argue here
that behavior is best explained by constellations of preferences that are
in turn rooted in other factors. Many of these foundations are structural. The view most widely held by international relations theorists,
for example, is that the global distribution of power is the structural
characteristic that determines the nature of global order. One currently
popular proposition links global predominance to stability; in particular
it links a hegemonic distribution of power to open international economic regimes.[49] Most blithely tie the distribution of power to the nature of the economic order, but few make the explicit causal argument
that depends on deducing a set of interests from a particular distribution of power and then ascertaining what order will emerge given
power and interests.[50] The argument here is that interests determine
regimes, and that the distribution of power should be viewed as one
determinant of interests and therefore of regimes. In other words, a
state's degree of power in the international system is one of the things
that explains its preferences, and the distribution of power between
states determines the context of interaction and the preference orderings of the interacting states and thus the incentives and prospects for
international regimes. Structural arguments should be recognized as
constituting the determinants of those different patterns of interest that
underlie the regimes themselves.

A similar structural argument can be used to explain subsystemic
regimes, for the extraregional context or structure can determine the
constellation of preferences among intraregional actors. Great powers
can often structure the choices and preferences of minor powers and
thus shape regional outcomes. Many of the cooperative arrangements
between Western European states immediately following the Second
World War can be said to reflect the way in which, through carrot and
stick, the United States structured the choices and preferences of those
states. The prisoners' dilemma also illustrates this, for the dilemma can
be seen as a parable of domination in which the district attorney structures the situation to be a dilemma for the prisoners.[51] Divide-and-

49. Recent exponents of the predominance model of stability, as opposed to the classical balance-of-power model of stability, include A. F. K. Organski, *World Politics*, 2d ed.
(New York: Knopf, 1968), pp. 338–76, and George Modelski, "The Long Cycle of Global
Politics and the Nation-State," *Comparative Studies in Society and History* 20 (April 1978):
214–35. The international political economy variant of the argument is provided by Stephen D. Krasner, "State Power and the Structure of International Trade," *World Politics*
28 (April 1976): 317–47; see also Stein, "The Hegemon's Dilemma."
50. This is precisely the way in which Krasner develops his argument in "State Power."
51. Tom Burns and Walter Buckley, "The Prisoners' Dilemma Game as a System of
Social Domination," *Journal of Peace Research* 11 (1974): 221–28.

conquer is one strategy by which the powerful can structure the interactions between others by determining for them their preferences
among a given set of choices.

There are other structural factors, such as the nature of technology,
and the nature of knowledge, that also determine actors' preferences
and thus the prospects for regimes. The nature of technology is critically important to a state's decision whether or not to procure weapons.
Typically, scholars have argued that states confront a security dilemma
in which they have prisoners' dilemma preferences. All states have a
dominant strategy of arming themselves, but all find the armed world
that results less preferable than a totally disarmed one. Yet the security
dilemma presumes either that offensive weapons exist and are superior
to defensive ones or that weapons systems are not easily distinguish-
able.[52] If only defensive weapons existed, however, no security dilemma
could arise. The actors would no longer have dominant strategies of
arming themselves, for the arms could not be used to exploit those who
had not armed, and procurement would not be a required defense
against exploitation at the hands of others' defensive weapons. The
interactions between states would no longer lead to a Pareto-deficient
equilibrium outcome, and there would be no need for an arms regime.
Thus the different constellations of preferences that exist in different
areas and create different incentives and prospects for international
regimes are in part a function of the nature of technology.

Changes in knowledge—the nature of human understanding about
how the world works—can also transform state interests and therefore
the prospects for international cooperation and regime formation. As
late as the middle of the nineteenth century there was enormous varia-
tion in national quarantine regulations, for example. As long as there
was no agreed body of validated knowledge about the causes of com-
municable disease and the nature of its transmission and cure, state
policy could and did reflect political concerns. Regulations to exclude
and isolate goods and individuals, ostensibly for health reasons, were
used as instruments for international competition and became the basis of conflict. But new medical discoveries—about the microbes that
cause such diseases as cholera and leprosy, about the transmission of
yellow fever by mosquitoes and plague by rat fleas, and of preventive
vaccines such as the one for cholera—transformed this situation by
providing a scientific foundation for new international agreements on
quarantine rules.[53] New knowledge thus changed states' preferences

52. Jervis, "Cooperation under the Security Dilemma."
53. The examples are from Charles O. Pannenborg, *A New International Health Order:*

and provided the basis for international cooperation and the depoliti-
cization of health care policy.[54]

   Just as structural factors underpin actors' preferences, so do internal
national characteristics. The interests of domestic economic sectors, for
example, can be the basis for national interests.[55] Even if a state's interests
do not reflect those of any specific sector or class, they may emerge from
a state's attributes. Large populations and high technology generate demands that will require a state to go abroad for resources if domestic access to them is inadequate.[56] Yet needed resources can be obtained by
exchange as well as by plunder. One cannot, therefore, move from a de-
lineation of internal characteristics to state behavior without incorporating some aspect of a state's relations and interactions with others.
Internal characteristics may determine a single actor's preferences, but
to ascertain outcomes, it is also necessary to know the interests of other
actors and to have a sense of the likely pattern of strategic interaction.[57]

Regime Maintenance and Change

The same factors that explain regime formation also explain regime
maintenance, change, and dissolution. Regimes are maintained as long

An Inquiry into the International Relations of World Health and Medical Care (German-
town, Md.: Sijthoff and Noordhoff, 1979), pp. 179–80.

   54. For a discussion of the role of scientific advances in the development of international health agreements, see Richard N. Cooper, "International Cooperation in Public
Health as a Prologue to Macroeconomic Cooperation," in *Can Nations Agree?: Issues in
International Economic Cooperation*, ed. Richard N. Cooper, Barry Eichengreen, C. Randall Henning, Gerald Holtham, and Robert D. Putnam (Washington, D.C.: Brookings Institution, 1989), pp. 178–254. Note that scientifically based health regulations can still
become the basis for political disagreement, as demonstrated by the Japanese response
to California's medfly spraying in 1981.

   55. See, for example, Peter Alexis Gourevitch, "International Trade, Domestic Coalitions, and Liberty: The Crisis of 1873–1896," *Journal of Interdisciplinary History* 8 (Autumn
1977): 281–313; and James R. Kurth, "The Creation and Destruction of International Regimes: The Impact of the World Market," paper delivered at the American Political Science
Association Meeting, Washington, D.C., August 1980.

   56. Robert C. North, "Toward a Framework for the Analysis of Scarcity and Conflict,"
International Studies Quarterly 21 (December 1977): 569–91.

   57. Note that this clearly distinguishes domestic sectoral from international structural
approaches. Although both approaches can be seen as delineating the determinants of
actor preferences, the international structural perspective can be claimed to determine
the constellation of all actors' preferences. Thus the existence of offensive weapons creates a prisoners' dilemma situation for any pair of nations. On the other hand, the sectoral
approach explains one actor's preferences at a time and so must be linked with an
analysis of the interaction between actors to explain outcome. This is, of course, why the
analysis of foreign policy is not equivalent to the analysis of international relations. Thus
the works of Allison, Gourevitch, Katzenstein, and Kurth, among others, which explain
foreign policy by reference to domestic economic or bureaucratic interests, remain incomplete precisely because they do not incorporate relations between nations. See also
the discussion in Chapter 7.

as the patterns of interest that gave rise to them remain. When these shift, the character of a regime may change; a regime may even dissolve entirely. Incorporating the determinants of interests leads one to argue that regimes are maintained only as long as the distribution of power (or the nature of technology or of knowledge, etc.) that determines a given constellation of interests remains. When the international distribution of power shifts, affecting, in turn, the preferences of actors, then the regime will change. Those who make a direct link between structure and regimes necessarily conclude that changes in the distribution of power lead to regime change. My argument here is more subtle. If interests intervene between structure and regimes, only those structural changes that affect patterns of interest will affect regimes. Further, since other factors also affect interests, it may be that the impact of changing power distributions on actors' preferences can be negated by other structural changes, such as those in technology. Or changes in the other factors, such as knowledge, can lead to regime change without a change in the distribution of power. This describes the history of quarantine regulations, for example. Together, these might explain why some changes in the distribution of power have clearly been linked with regime changes whereas others have not.[58]

Regimes may be maintained even after shifts in the interests that gave rise to them, however. There are a number of reasons for this. First, nations do not continually calculate their interactions and transactions.[59] That is, nations reassess only periodically their interests and power or the institutional arrangements that have been created to deal with a particular configuration of them. Once in place, the institutions serve to guide patterned behavior, and the costs of continual recalculation are avoided. Decision costs are high, and once paid in the context of creating institutions, they are not continually borne.[60]

An alternative argument is that the legitimacy of international institutions emerges not from any waiving of national interest but from an interest developed in the institutions themselves. Any shift in interests does not automatically lead to changes in the regime or to its destruc-

58. The recognition of the multiple determination of actors' interests also makes possible an issue approach to international politics that is not necessarily issue-structural.

59. Thorsten Veblen ridiculed the concept of marginal utility, writing, "It is not conceivable that the institutional fabric would last overnight," if all exchange relationships "were subject to such a perpetual rationalized, calculating revision, so that each article of usage, appreciation, or procedure must approve itself de novo." Quoted by Robert Kuttner, "The Economist's Heart," *The New Republic*, October 2, 1989, p. 39.

60. One can, of course, expect there to be lags between changes in interests and actors' behavior; see Michael Nicholson, *Oligopoly and Conflict: A Dynamic Approach* (Liverpool: Liverpool University Press, 1972). Schick distinguishes realization lags from adaptation lags in "Some Notes on Thinking Ahead," p. 790.

tion, because there may well be uncertainty about the permanence of the observed changes. The institutions may be required again in the future, and destroying them because of short-term changes may be very costly in the long run. Institutional maintenance is not, then, a function of a waiving of calculation; it becomes a factor in the decision calculus that keeps short-term calculations from becoming decisive.[61] Because international institutions involve sunk costs, they are not likely to be readily changed or destroyed. The costs of reconstruction are likely to be much higher once regimes are consciously destroyed. Their very existence changes actors' incentives and opportunities.[62]

There is, however, an alternative to the explanation that maintenance of regimes is merely a perpetuation of the exogenous factors that occasioned their rise. It may be that neither sunk costs nor delays in recalculation or reassessment are responsible for regime maintenance. Max Weber argues that tradition provides legitimacy and is one basis for the maintenance of a political order, and this argument can be extended to international relations. International regimes can be maintained and sustained by tradition and legitimacy. Even those international institutions that exist in an anarchic environment can attain legitimacy that maintains patterned international behavior long after the original basis for those institutions has disappeared. Thus, even though the constellations of interest that give rise to regimes may change, the regimes themselves may remain. This circumstance can be explained by means of interests by arguing that actors attach some value to reputation and that they damage their reputations by breaking with customary (i.e., traditional) behavior.[63] An actor that comes to prefer independent decision making to the maintenance of the regime may nevertheless stick with the latter because it values an undiminished reputation more than whatever it believes it would gain by departing from the established order.

Finally, there is a possibility that the creation of international regimes leads not to the abandonment of national calculation but to a shift in

61. The contrasting implications of long-term and short-term calculations are discussed in Chapter 4.

62. One can argue that regimes actually change actors' preferences. The property rights argument about dealing with externalities through changes in liability rules is an example of a situation in which prearranged agreements are specifically devised in order to change utilities in subsequent interaction; see John A. C. Conybeare, "International Organization and the Theory of Property Rights," *International Organization* 34 (Summer 1980): 307–34.

63. George A. Akerlof, "A Theory of Social Custom, of Which Unemployment May Be One Consequence," *Quarterly Journal of Economics* 94 (June 1980): 749–75.

the criteria by which decisions are made. Institutions created to ensure international coordination or collaboration can themselves serve to shift decision criteria and thus lead nations to consider others' interests in addition to their own when they make decisions. Once nations begin to coordinate their behavior and, even more so, once they have collaborated, they may become joint-maximizers rather than self-maximizers. The institutionalization of coordination and collaboration can become a restraint on individualism and lead actors to recognize the importance of joint maximization. Those that previously agreed to bind themselves out of self-interest may come to accept joint interests as an imperative. This may be especially true of collaboration regimes, precisely because they require that actors trust one another not to cheat even though they all have an incentive to do so. In these situations one nation's leaders may come to have an interest in maintaining another nation's leaders in power, for they have worked together to achieve the optimal nonequilibrium outcome and they trust one another not to cheat. Recognition of the importance of maintaining the position of others may become the basis for the emergence of joint maximization as a decision criterion for actors.[64]

Conclusion

The problems of analyzing regime formation, maintenance, and dissolution demonstrate the clear necessity for a strategic-interaction approach to international politics. State behavior does not derive exclusively from structural factors like the distribution of power; neither can such behavior be explained solely by reference to domestic sectors and interests. Structure and sectors play a role in determining the constellation of actors' preferences, but structural and sectoral approaches are both incomplete and must be supplemented by an emphasis on strategic interaction between states. It is the combination of actors' preferences and the interactions that result from them that determine outcome, and only by understanding both is it possible to analyze and understand the nature of regimes in an anarchic world.

We have long understood that anarchy in the international arena does not entail continual chaos; cooperative international arrangements do exist. This chapter differentiates the independent decision making that characterizes "anarchic" international politics from the joint decision making that constitutes regimes. In doing so, it distin-

64. See Chapter 6.

guishes the natural cooperation that results from harmonious interests from those particular forms of collective decision making that define regimes. Sovereign nations have a rational incentive to develop processes for making joint decisions when confronting dilemmas of common interests or common aversions. In these contexts, self-interested actors rationally forgo independent decision making and construct regimes.

The existence of regimes is fully consistent with both realist and liberal views of international politics, in which states are seen as sovereign and self-reliant. It is the very autonomy of states and their self-interests that lead them to create regimes when confronting dilemmas.

3

Misperception and
Strategic Choice

If actors behave purposively given the information available to them,
perception—the information that actors possess about others—can be
a critical determinant of behavior. At times, therefore, perception, and
hence misperception, can provide the foundation for the particular
choice between cooperation and conflict.[1]

Scholars have long assumed, in fact, that misperception leads nations
to enter conflicts they would otherwise avoid, and have attributed many
wars, including both world wars and the Cold War, at least in part, to
misperception.[2] This belief is equivalent to holding that full knowledge
leads to cooperation.

The view that conflict emerges from misunderstanding, miscom-
munication, and misperception is a central tenet of liberalism. Its obvious counterpoint suggests that conflicts could be avoided if people
and governments merely knew and understood one another. Those
who link economic interdependence with international cooperation,

1. This is a much expanded and somewhat altered version of my article "When Misperception Matters," *World Politics* 34 (July 1982).
2. Ralph K. White, "Misperception as a Cause of Two World Wars," in *Nobody Wanted
War: Misperception in Vietnam and Other Wars*, rev. ed. (Garden City, N.Y.: Doubleday,
1970), pp. 3–33; Chihiro Hosoya, "Miscalculations in Deterrent Policy: Japanese-U.S. Relations, 1938–41," *Journal of Peace Research* 5 (1968): 97–115; Stanley Hoffmann, "Revi-
sionism Revisited," in *Reflections on the Cold War: A Quarter Century of American Foreign
Policy*, ed. Lynn H. Miller and Ronald W. Pruessen (Philadelphia: Temple University Press,
1974), pp. 3–26; and Ralph K. White, "Misperception in the Arab-Israeli Conflict," *Journal
of Social Issues* 33 (Winter 1977): 190–221.

for example, argue that the very existence of exchange promotes peace by developing shared interests among peoples.[3]

The hope that improved prospects for peace would evolve from communication and interaction received heightened attention in the nineteenth century when revolutions in the means of communication and travel sped and expanded the scope of international contact. Railroads, steamships, and telegraph lines increasingly linked far parts of the world. These changes all seemed to liberals to promise improved prospects for peace.

Many recommendations flowed from these assumptions about the ramifications of improved communication and increased understanding among peoples. Both government policies and private programs tried to expand the scope of interaction between the citizens of different countries. The nineteenth century saw the convocation of conferences of every kind, the proliferation of international exhibitions and world fairs, and, in 1896, the establishment of the modern Olympic games.

So strong was the belief in the power of contact to prevent hostilites that not even the outbreak of World War I discredited the notion that these proliferating meetings could encourage peace. On the contrary, some proponents of interaction blamed not the inefficacy of contact, but the lack of communication between the leaders of the disputing nations. Giving the leaders more opportunity to discuss their concerns, this logic held, would have made the difference. The solution for the future would be to establish a mechanism to ensure such talks, and the League of Nations was born.

This faith in communication as a means to understanding and, therefore, to cooperation continues to the present day. Although the League could not stop World War II, a new organization, the United Nations, was established to facilitate international dialogue. The postwar era has been replete with institutionalized and ad hoc summits intended to bring national leaders face-to-face. Not only political contact among governments, but cultural and scientific exchanges among peoples have also been promoted.[4]

3. This was the view of Richard Cobden, among others. The present discussion of liberalism and international communication draws upon my article "Governments, Economic Interdependence, and International Cooperation," in *Behavior, Society, and Nuclear War*, vol. 3, ed. Philip E. Tetlock, Jo L. Husbands, Robert Jervis, Paul C. Stern, and Charles Tilly (New York: Oxford University Press, for the National Research Council of the National Academy of Sciences, forthcoming).

4. Karl Deutsch argues that interaction does more than improve understanding, that it also generates community. He believes it essential to the development of a group

Two key problems attend an emphasis on the direct value of inter-actions, however. There is no logical reason to expect knowledge and familiarity either to generate common interests or to reduce conflicts of interest. If familiarity and knowledge were at the heart of cooperation, families would not feud, couples would not divorce, and war would not be most common among states that share borders.[5] Some inter-action is required for conflicts of interest to occur. Just as actors who do not interact cannot cooperate, so they cannot fight. Knowledge of others' needs is the basis both for empathy and for extortion and exploitation.[6]

The importance of misperception, on the other hand, has been em-phasized by the literature on the role of cognitive processes and by analyses of decision making in the study of foreign policy. Yet a theory of misperception remains to be formulated. The most definitive work on the subject, *Perception and Misperception in International Politics* by Robert Jervis, presents a categorization of types of misperception and provides illustrations for each.[7] But the final chapter of the book is entitled "In Lieu of Conclusions," and we still do not know what misperceptions occur, under what conditions, and with what conse-quences. All too often, the mere occurrence of misperception is taken as prima facie evidence that it affected the misperceiving actor's deci-

consciousness and argues that the extent of social interaction defines the bounds of the community. Growing international contact would lead not merely to international co-operation, he holds, but to the emergence of a new, integrated community of nations. Deutsch's own work focuses on a range of issues: the role of communications, transac-tions in general, and trade in particular; see his "Power and Communication in Inter-national Society," in *Conflict in Society*, ed. Anthony de Reuck and Julie Knight (Boston: Little, Brown, 1966), pp. 300–316; "The Impact of Communications upon Theory of Inter-national Relations," in *Theory of International Relations: The Crisis of Relevance*, ed. Abdul A. Said (Englewood Cliffs, N.J.: Prentice-Hall, 1968), pp. 74–92; "The Propensity to Inter-national Transactions," *Political Studies* 8 (1960): 147-56; "Transaction Flows as Indicators of Political Cohesion," in *The Integration of Political Communities*, ed. Philip E. Jacob and James V. Toscano (Philadelphia: Lippincott, 1964), pp. 75–97; Karl W. Deutsch, Chester I. Bliss, and Alexander Eckstein, "Population, Sovereignty and the Share of Foreign Trade," *Economic Development and Cultural Change* 10 (1962): 353-66; and Karl W. Deutsch and Alexander Eckstein, "National Industrialization and the Declining Share of the Interna-tional Economic Sector, 1890-1959," *World Politics* 13 (1961): 267-99.

5. David Wilkinson, *Deadly Quarrels: Lewis F. Richardson and the Statistical Study of War* (Berkeley and Los Angeles: University of California Press, 1980), chap. 5.

6. Familiarity can also breed contempt. To know more about others is not always to like them better. And even understanding another's position need not lead to sympathy for it. Karl Deutsch is aware of this. Thus he sometimes emphasizes that there must be a multiplicity of interactions and value compatibility; see Deutsch, *Nationalism and Its Alternatives* (New York: Knopf, 1969), pp. 103-4.

7. Robert Jervis, *Perception and Misperception in International Politics* (Princeton, N.J.: Princeton University Press, 1976).

sion and thus the outcome itself. Moreover, it is universally suggested that the result of misperception is conflict that would otherwise have been avoidable. Although international conflicts are often attributed to misperception, international cooperation never is.

This chapter is a first step toward a theory of misperception. I begin by distinguishing between different kinds of misperception and between misperceptions and miscalculations. I then assess the circumstances in which misperception is a determinant of choice and outcome. The focus here is on one actor's misperception of another's preferences.[8] My conclusions suggest that misperception does not always affect an actor's choices or determine outcome; that when misperception does have such effects, it is in a narrow range of circumstances; and that misperception can lead to cooperation as well as to conflict. Further, I elucidate the assumptions about international relations that are implicit in any emphasis on the role of misperception in international politics.

8. I emphasize the misperception of others' preferences as opposed to the misperception of their intentions. I draw the distinction in the hope that it will help those who infer intention from outcome and those who identify the value of payoffs (positive or negative) with an actor's motivations.

My point can best be explicated with an example. In 1955 the Soviet Union signaled its desire to negotiate the withdrawal of the superpowers from Austria (for a discussion of the treaty, see Deborah Welch Larson, "Crisis Prevention and the Austrian State Treaty," *International Organization* 41 [Winter 1987]: 27-60). An agreement to accomplish this, the Austrian State Treaty, was quickly reached. The payoffs to the West were clearly positive. Nonetheless, a question of interpretation remains. Perhaps Soviet intentions were not benign; perhaps they took this step in the hope that one consequent outcome would be that West Germany would remain unarmed. International relations theorists would typically say that Soviet intentions were unknown. For some, the intentions are critical in an assessment of the implications of misperception. They are not necessarily important, however.

Soviet intentions, hopes, or expectations did not matter. What would follow the superpower withdrawal and establishment of Austrian neutrality did matter. Either Germany would or would not rearm. If it did, the outcome for the Soviets would be negative even if they never dreamed that the treaty might lead to such a result. Similarly, Soviet intentions would not in any way affect the value of the outcome for the United States and the Western alliance. It would be positive or negative irrespective of Soviet desires or motivations.

The point is that the potential payoffs associated with the treaty entailed multiple components. The immediate payoff involved Austrian neutrality. But there were also payoffs associated with the consequences that the treaty would generate. Such second-order payoffs can be fully unpredictable, although the actors may have hopes or suspicions about what will happen. If so, they will assess these expected consequences as well as the immediate outcome when deciding what to do. But this assessment is independent of others' intentions. They may wish it so, but that does not make it so.

For a brief general discussion of problems in assessing intentions, see R. B. Zajonc, "Altruism, Envy, Competitiveness, and the Common Good," in *Cooperation and Helping Behavior: Theories and Research*, ed. Valerian J. Derlega and Janusz Grzelak (New York: Academic Press, 1982), pp. 417-36.

Misperception of Capabilities and Intentions

A wide range of phenomena have been grouped under the general rubric of "misperception," which attends much social interaction and can be a determinant of choice. One author even admits that his study conflates misperception with misconception and defines "misperception" so broadly as to include "even the individual's most basic assumptions about the nature of the world and of man."[9] To assess the implications of misperception for international relations requires, however, that it be more narrowly and precisely defined and that misperceptions of capabilities be distinguished from misperceptions of intentions, preferences, and interests.[10]

The relative ease of identifying and measuring already deployed military power, the most obvious manifestation of capability, has meant that a nation's debates over foreign and military policy typically concentrate on its opponent's intentions.[11] This has become ever more the case as the sophistication of the technical means available for counting weapons has grown. Questions about capability have increasingly centered less on the other country's current stockpiles and more on the items it is procuring or hopes to deploy. To answer these questions, analysts combine solid information about the current status quo with informed judgments about ongoing or planned acquisitions.

In short, debates about capability must necessarily include some assessment of intentions. In the middle 1950s, for example, U.S. analysts combined solid evidence of the Soviet Union's ability to produce bombers at a certain rate with the assumption that it would produce as many as possible and thereby predicted a bomber gap—suggesting that the United States would shortly fall behind in the number of deployed bombers.[12] Eventually the Cassandras scaled down their dire predic-

9. White, *Nobody Wanted War*, p. 7.
10. For a review, see Jack S. Levy, "Misperception and the Causes of War: Theoretical Linkages and Analytical Problems," *World Politics* 36 (October 1983): 76–99. Note that preferences reflect an assessment of capabilities.
11. Intelligence bureaus were quite good at assessing capabilities prior to both world wars; see Ernest R. May, ed., *Knowing One's Enemies: Intelligence Assessment before the Two World Wars* (Princeton, N.J.: Princeton University Press, 1986). Yet they still inaccurately assessed military doctrines and the course of a prospective war. Although certain aspects of power, such as morale and commitment, have also been correctly assessed before the fact, they remain an especially problematic area for forecasters. The ability to mobilize resources for an extended war is not a fixed characteristic but is related to the nature of the outbreak of war; see Arthur A. Stein, *The Nation at War* (Baltimore: Johns Hopkins University Press, 1980).
12. Allen Dulles, *The Craft of Intelligence* (1963; Boulder, Colo.: Westview Press, 1985),

tions, but the forecasts of a bomber gap rapidly gave way to those of a missile gap as Americans reacted with horror to the Soviet launch of the world's first earth satellite, Sputnik I, on October 4, 1957. The USSR led the United States in advanced technology, and Sputnik signaled its ability to develop intercontinental ballistic missiles. As a result, the U.S. intelligence debate turned to the question of when such a missile buildup would begin and how rapidly it would proceed, with analysts basing their forecasts on estimates of Soviet production rates.[13] Subsequent observations made it clear, however, that the Soviets never produced missiles anywhere near the predicted rate. Even when assessments of current capabilities are excellent, good forecasting requires that appraisals of intentions also prove accurate. Predictive judgments of future capability necessarily include assessments of intentions as well.[14]

The analysis of extant capabilities can also require the examination of intentions. In 1962, for example, the American determination of what weapons the Soviet Union was shipping to Cuba had to rest on the knowledge of what it could send, the number and size of its shipments, and a conjecture as to whether the cargo included ground-to-ground missiles. Once deployment began, no one questioned its ongoing occurrence. Before the fact, however, judgments about the prospects for deployment had to be based on an assessment of Soviet plans.[15]

Misperception and Miscalculation

Predicting capability is like making any other kind of forecast. It involves a probabilistic assessment of an uncertain and unknown future. Picking a suitable military policy is not unlike selecting an appropriate economic one—both are necessarily based on educated guesses about

---

pp. 162–63, and Colin S. Gray, "Gap Prediction and America's Defense: Arms Race Behavior in the Eisenhower Years," *Orbis* 16 (Spring 1972): 257–74.

13. Lawrence Freedman, *U.S. Intelligence and the Soviet Strategic Threat* (Boulder, Colo.: Westview Press, 1977), chap. 4; Edgar M. Bottome, *The Missile Gap: A Study of the Formulation of Military and Political Policy* (Rutherford, N.J.: Fairleigh Dickinson Press, 1971); James C. Dick, "The Strategic Arms Race, 1957–1961: Who Opened a Missile Gap?" *Journal of Politics* 34 (November 1972): 1062–1110; Roy E. Licklider, "The Missile Gap Controversy," *Political Science Quarterly* 85 (December 1970): 600–615.

14. Freedman, *U.S. Intelligence and the Soviet Strategic Threat*, p. 184, puts it the following way: "The conventional distinction between 'estimates of capabilities' and 'estimates of intentions' breaks down in practice."

15. In some cases, of course, spies or satellites may be able to provide documentary evidence about intentions or evidence of ongoing deployments or of military shipments en route.

what is going to be. In both cases, knowing the odds of a certain event's coming to pass is not the same as knowing that it will or will not happen. High-probability events still fail to occur, and low-probability events sometimes do, even if infrequently. Long shots win, though not often, and "sure" things fail to materialize.[16]

People must often make decisions without knowing for certain what the future holds, with their knowing at best the odds of different outcomes. A priori and a posteriori odds almost always diverge, however. After the fact (a posteriori), the odds of something's having happened are either zero or one hundred percent, for either it has occurred, or it has not. Hindsight always provides 20/20 vision. Yet before the fact (a priori), when the outcome is still unknown and uncertain, the odds are neither zero nor one hundred, but somewhere in between. Anyone who puts the odds at either extreme, although wrong before the fact, will nevertheless turn out to be fully right or completely wrong.

President Kennedy said at one point during the Cuban missile crisis of 1962 that he believed the chance of war to be between 1 in 3 and 1 in 2.[17] Afterward, of course, the probability was zero.[18] The president was certainly correct, however, when he declared that there existed some chance of war. Had Kennedy said that war would not occur, he would have been wrong in that there was a finite chance of war, but his forecast would have been more accurate than all those suggesting that there existed any chance at all. Nonetheless, scholars would have castigated his myopia and offered an array of psychological explanations for his inability to recognize and accept the reality that war was possible.

As long as the future is unknown, people remain at least somewhat uncertain about what to do, and the miscalculation that can result is endemic to human decisions. No one would choose an incorrect forecast over a correct one. But inherent uncertainty about the future can lead to bad predictions, miscalculations, and misassessments. Most economic exchanges of assets typically involve incongruent assessments of the future. Indeed, uncertainty about the future is necessary if markets are to function. Every stock sale involves a buyer and a seller, and in most cases they make different assessments about the future

16. A classic distinction is between uncertainty, where the probabilities are unknown, and risk, where they are known.

17. Theodore C. Sorensen, *Kennedy* (New York: Harper and Row, 1965), p. 705.

18. Even retrospectively, however, one can argue that the probability of war had been higher and that the world lucked out. As John Kenneth Galbraith put it, "We were in luck, but success in a lottery is no argument for lotteries." John Kenneth Galbraith, "The Plain Lessons of a Bad Decade," *Foreign Policy*, no. 1 (Winter 1970–71): 32.

price of the stock. After the fact, only one of the two actors can be said to have calculated correctly. On Black Thursday, the day most associated with the Great Crash of 1929, there were, in the words of John Kenneth Galbraith, not only "12,894,650 shares sold...precisely the same number were bought."[19] Yet that exchange would most likely not have taken place if both buyer and seller had known for certain what the price of the stock would be at the end of the day.

Miscalculation can also attend most contests in which there will be a winner. Teams that play in sports championship series, for example, have already demonstrated their skill in defeating rivals; both teams enter the competition believing they can win. Both may fully believe that they *will*. In this case the members of the losing team will have miscalculated—they did not win although they believed they would.[20]

Wars are no different. If every war involves at least two nations, each of which believes it can prevail, all but the eventual victor miscalculate in choosing to become belligerents.[21] As Geoffrey Blainey maintains, "Wars usually begin when two nations disagree on their relative strength, and wars usually cease when the fighting nations agree on their relative strength."[22]

In other words, one cause of war is that nations hold incongruent assessments of their relative power.[23] The state that initiates the fighting

19. John Kenneth Galbraith, *The Great Crash, 1929* (Boston: Houghton Mifflin, 1979), p. 109.

20. Even convergent forecasts may lead to dissimilar decisions. Bookmakers generate odds about sporting events in such a way as to equalize the betting on either side. Thus two individuals may agree on who will win but when informed of the betting line on a particular sports event may choose different sides of the bet.

21. Wars can also begin when states see the odds the same way but one chooses the likely choice and the other takes a long shot. Further, there may be residual cases in which states expect to snatch victory from the jaws of defeat, expecting net positive benefits even from a loss (a Fenwickian, or mouse-that-roared, strategy). Finally, states will sometimes resist invasion even without hope of victory.

22. Geoffrey Blainey, *The Causes of War* (New York: Free Press, 1973), p. 246. In a formal analysis by Dagobert L. Brito and Michael D. Intriligator, a voluntary redistribution of resources, rather than war, occurs when all nations are fully informed. See Brito and Intriligator, "Conflict, War, and Redistribution," *American Political Science Review* 79 (December 1985): 943–57. See also Donald Wittman, "How a War Ends: A Rational Model Approach," *Journal of Conflict Resolution* 23 (December 1979): 743–63. Wittman argues that war occurs only if at least one nation is excessively optimistic. The same is true of strikes. If labor and management both foresee the same outcome, there should be no strike. Knowing the outcome of the negotiations for a new contract, they would prefer merely to sign it without having to bear the cost of the work stoppage. Strikes occur because of divergent assessments. Much the same argument is made about international crises, that they occur only because of incomplete or imperfect information. See Robert Powell, "Crisis Bargaining, Escalation, and MAD," *American Political Science Review* 81 (September 1987): 717-35, and the list of citations he provides.

23. This is true whether one subscribes to the balance-of-power view that an equilib-

believes it can win; no state will become involved in hostilities it knows it can lose.[24] By the same token, the attacked state resists because its leaders believe that it can win. But one of the belligerents will lose. In Blainey's words, what is "so often unintentional about war [is]...not the decision to fight but the outcome."[25] On the eve of war, one nation miscalculates the odds of victory, and "in that sense every war comes from a misunderstanding. And in that sense every war is an accident."[26]

If the decision to go to war is based not on the presumption of victory but on the willingness to risk losing, miscalculation is not an issue, however. A state has not miscalculated if it loses a war it has entered in the belief that it has a sufficient chance of winning to make fighting worthwhile. However small the odds it calculates, they are high enough for it to start a war.[27] Defeat is undesirable but not fully unexpected.

Whether believing victory to be a long shot or a likelihood, the state that makes the decision to go to war must be totally dedicated to this endeavor. It is often difficult, however, to make such a commitment and still keep in mind that the outcome is uncertain. Hence, in Galbraith's words, "we compensate for our inability to foretell...by asserting positively just what the consequences will be."[28] That people generate enthusiasm for a course of action or even a sense of certainty about its still-future outcome does not mean, however, that they are

---

rium of power is conducive to peace or to the predominance argument that a disequilibrium of power is conducive to peace.

24. This is important in understanding the Japanese decision to attack Pearl Harbor in 1941. The Japanese knew that they would lose any prolonged war with the United States. They were aware of America's ability to mobilize greater resources and bring more power to bear in any extended contest. Thus it has become important to assess whether the Japanese decision to attack was rational or not. One answer is to point out that they knew they would lose a prolonged war but felt that there was a reasonable chance that the United States, not wanting to wage a protracted one, might negotiate more favorable terms after experiencing a loss such as that at Pearl Harbor. Moreover, waiting would only worsen the Japanese situation, which was deteriorating daily under the weight of America's oil embargo. In short, the Japanese did not know for certain that they would lose, although they certainly knew that the odds were against them. This case is discussed more fully in Chapter 5.

25. Blainey, *Causes of War*, p. 144.

26. Ibid., p. 145.

27. Chapter 5 includes a discussion of when states might be willing to take slim chances.

28. Galbraith, *Great Crash*, p. 171. Indeed, once people commit themselves to a course of action, they buoy themselves and bolster their own decisions. It often seems that saying "it will be so" makes it so. Galbraith points out that "by affirming solemnly that prosperity will continue, it is believed, one can help ensure that prosperity will in fact continue. Especially among businessmen the faith in the efficiency of such incantation is very great" (p. 16).

unaware of the odds while deciding what to do, or that they miscalculate or misperceive.

This need to make wholehearted commitments in the context of ambiguity is a challenge that confronts people often.[29] The college athlete of average talent, for example, faces a major dilemma. The odds of getting to play professionally are slim. But taking time away from sports in order to ensure an alternative career by concentrating on academic coursework almost invariably ensures the failure to become a professional athlete. Thus any who believe the small chance is worth taking must ignore the poor odds and devote themselves completely to their sports, even if that means having nothing to fall back on later.

By the same token, although a nation's leaders must make decisions to wage war on the basis of probabilities, they cannot publicly display any doubts about their chosen course of action.[30] Once they have decided to act, they must commit themselves and their populations fully, for the failure to do so may actually ensure defeat. Hence belligerents generally display confidence, optimism, and even ebullience on the eve of war. Leaders' proclamations of certain victory can thus reflect either the expectation of victory that accompanies the original decision to act or the need to unify their nation behind this chosen course—or both.[31]

Decisions in an uncertain world can entail divergent forecasts, self-encouragement, and even miscalculation, yet none of these necessarily reflects misperception. Divergent forecasts are to be expected when the future is unknown. Accurate predictions can be based on misperceptions, and inaccurate predictions can be based on valid perceptions (most economic forecasts probably fall into this category). Neither does self-emboldening behavior signal the existence of misperception. Indeed it may reflect an all too functional delusion necessary to ensure success and absolutely essential to make long shots pay off.[32] Evidence of psyching oneself up is not necessarily a sign of misperception. Misperception does not involve erroneously predicting the future but inaccurately seeing the present. It occurs when actors misread available information; neither incorrect predictions nor bluster provide evidence either of misperception or that misperception has determined choice.

29. This is an essential human ability. Patients, for example, decide on courses of treatment in the context of morbidity data provided by their doctors. Having chosen, they have confidence in what they have done; often, that confidence itself enhances their ability to survive.

30. Herein lies the similarity of FDR, Churchill, and Reagan, in stark contrast to Carter, who wore his insecurities on his sleeve.

31. Such proclamations can also, of course, reflect dissonance reduction.

32. See the discussion in Chapter 7.

Choices, Interests, Expectations, and Misperceptions

Misperceiving another's preferences, interests, or intentions most obviously matters when an actor chooses a course of action in order to affect another party. It is a truism, for example, that one must perceive another's wishes accurately in order to behave in accordance with them. But for self-interested individualists concerned with maximizing their own returns, misperception matters only under certain circumstances.

Misperception matters only in relations between interdependent actors. If states are independent of one another, in the sense that one's decisions do not affect the other's payoffs, then misperception is irrelevant. To borrow the analogy often used by international systems theorists, misperception is irrelevant if world politics approximates a competitive market in which states act as firms. In such a world the decisions of one actor do not affect the payoffs of others, and an assessment of the preferences of others is unnecessary.[33] If, however, world politics revolves primarily around a few major powers and approximates an oligopolistic market with imperfect competition, the actors can be seen as interdependent in the sense that the actions of any state affect those of others.

Finally, the belief that misperception is important necessarily implies that international politics is a variable-sum game. In any constant-sum game an actor can determine another's preference ordering simply by recognizing the game as constant-sum and knowing its own preferences.[34] If the two actors' payoffs add up to the same constant in each of the four possible outcomes, one actor's worst outcome must be the other's best, and so on. In other words, misperception, the incorrect

33. In a competitive market, an actor can misperceive the market and the nature of supply-and-demand conditions. But this does not involve misperception of any individual actor's preferences. The market analogy is prevalent in the works of Kenneth N. Waltz and Morton A. Kaplan. See Waltz, *Theory of International Politics* (Reading, Mass.: Addison-Wesley, 1979); and Kaplan, *Towards Professionalism in International Theory: Macrosystem Analysis* (New York: Free Press, 1979).

34. Actors may, of course, not recognize that the game is constant-sum. But some issues, such as territory, are by nature constant-sum (also see Chapter 5). If an actor not only knows its own preferences and the fact that it is a constant-sum game but recognizes the game as zero-sum, it can determine not only the other's preferences but its actual utilities as well. Finally, there is the possibility, even in constant-sum games, of misperceiving chance events that can intercede. See the discussion of poker in John von Neumann and Oskar Morgenstern, *Theory of Games and Economic Behavior* (Princeton, N.J.: Princeton University Press, 1944).

assessment of another's preferences, cannot occur in situations of pure conflict in which one actor's gain comes at another's expense. Misperception cannot matter in such situations; it can matter only in relationships that combine elements of cooperation and conflict.

The argument that misperception affects an actor's decision presumes that the actor has a choice. Some scholars maintain that actors do not always see that they have a choice, and misperceive others to have a wider latitude.[35] In fact, national leaders may not have choices because of structural or systemic constraints, or because of their own cognitive processes. But if they see themselves as having only a single course of action, their assessment of the preferences of others is moot, and their belief that others have alternative choices affects only their expectations. There is no reason to argue that they would have acted differently had they perceived others accurately.

Misperception need not affect the decision of an interdependent actor who does have a choice, even when it affects that actor's expectations. It would not, for example, affect the behavior of any actor with a "dominant strategy"—a course of action that maximizes its returns no matter what others do. Such an actor need not know or care about the preferences of others; at most, an inaccurate assessment (misperception) of another's preferences will affect its expectations.[36]

The American-Japanese crisis of 1940–41 illustrates the effect of misperception on expectations. In a study of this and other crises, Glenn H. Snyder and Paul Diesing follow the convention of dichotomizing the choices of states as cooperation/concession and defection/standing firm.[37] Figure 9 illustrates the interaction between two states with those two choices, an interaction that will result in one of four different outcomes: either both actors cooperate (CC), both defect (DD), or one defects while the other cooperates (CD and DC).

35. L. L. Farrar, Jr., "The Limits of Choice: July 1914 Reconsidered," *Journal of Conflict Resolution* 16 (March 1972): 1–23; and Ole R. Holsti, *Crisis Escalation War* (Montreal: McGill-Queen's University Press, 1972).

36. Note also that a dominant strategy implies the simultaneous existence of different intentions or motivations. Defection in a prisoners' dilemma game, for example, reflects both an actor's temptation to obtain the best outcome by taking advantage of another's cooperation and its fear of being exploited and obtaining the sucker's payoff that results from cooperating when the other defects. There is no way to pull these motivations apart. Nevertheless, the Athenians unabashedly ranked the reasons for the expansion of their empire as being "chiefly for fear, next for honour, and lastly for profit." See Donald W. Hanson, "Thomas Hobbes's 'Highway to Peace,' " *International Organization* 38 (Spring 1984): 329-54, specifically p. 339.

37. Glenn H. Snyder and Paul Diesing, *Conflict among Nations: Bargaining, Decision Making, and System Structure in International Crises* (Princeton, N.J.: Princeton University Press, 1977).

Figure 9. A general two-actor game

	Actor B	
	C: Accommodative strategy (B concedes or cooperates)	D: Coerceive strategy (B defects, stands firm, makes no concessions)
C: Accommodative strategy (A concedes or cooperates)	CC Mutual cooperation or compromise	CD B gets its way
D: Coercive strategy (A defects, stands firm, makes no concessions)	DC A gets its way	DD Mutual defection, deadlock, or war

(Actor A labels the rows)

Snyder and Diesing designate the American-Japanese crisis of 1940–41 as a game of "deadlock" (Figure 10) in which both the Japanese and the Americans had dominant strategies. In addition, both most preferred to stand firm while the other capitulated. Next, each preferred war (both standing firm); then, mutual compromise. The least desirable outcome for each was to capitulate while the other did not. The outcome that emerged from their independent choices, made in line with their dominant strategies of standing firm, was mutual defection. This outcome was an "equilibrium" one—that is, an outcome from which no individual actor can shift unilaterally without making itself worse off.

If the United States and Japan had each accurately perceived the other's preferences, each country would have expected war. Each might

Figure 10. Deadlock

	Actor B	
	Cooperate	Defect*
Cooperate	2, 2	1, 4
Defect*	4, 1	3, 3**

(Actor A labels the rows)

*Actor's dominant strategy
**Equilibrium outcome

have believed, however, that the other had a dominant strategy not of defection but of cooperation. In this case each would have expected the other's capitulation. Alternatively, each might have believed that the other's strategy was not dominant but contingent. The United States, for example, might have believed that the Japanese would stand firm only if it capitulated, but that they were certain to compromise if the United States stood firm. But because each knew that it would stand firm no matter what the other did, each could have expected only war or capitulation on the other's part. Either side's misperception of the other's preferences would lead it to anticipate an outcome other than that which actually occurred. Because misperception can affect only expectations and not decisions, both sides may have been surprised to find themselves at war. But they would not have acted differently had they perceived the other's preferences accurately.[38]

This distinction—that misperception can lead an actor with a dominant strategy to expect an outcome different from the actual one without affecting its own course of action[39]—is an important one, especially in view of the recent interest in strategic surprise. Evidence of surprise, or of inaccurate expectations, cannot be used to infer that an actor would have acted differently had it perceived the other's preferences accurately. Many examples of strategic surprise are cases in which the actors were clearly enemies and in which the surprised state was supposedly vigilant in its continual assessment of its opponent. That it was, in fact, surprised is only evidence that misperception affected its expectations. One certainly would not conclude that it was misperception that led to a conflict that would otherwise have been avoidable.

The Soviet invasion of Afghanistan in 1979 provides a recent illustra-

38. This analysis provides an interesting perspective on the question of whether President Roosevelt acted to bring on war with Japan. It is not accurate to suggest that he wanted war, for the most preferred outcome was Japanese capitulation to American demands. On the other hand, he did prefer war to either compromise or capitulation by the United States, and he understood that the American commitment to standing firm made war a possible outcome. In view of his belief that Japan would capitulate, Roosevelt did not expect war to be the outcome, but he did intend to stand firm, the possibility of war notwithstanding.

An interesting revisionist challenge might then be the following: Roosevelt did not misperceive Japanese preferences, recognized the game to be "deadlock," and knew that war was coming. Although he continued to stand firm by American demands, he did not actively prepare for war because he did not want to be accused of wanting war or of bringing it about. He thus pretended to expect Japanese concessions, knowing full well that, given the context, war was inevitable.

39. Misperception need not even affect an actor's expectations. An actor can assess the actual preference ordering inaccurately and still perceive another's dominant strategy accurately.

tion. By all accounts, the Soviets were quite understandably surprised by the U.S. reaction. Although they did not expect plaudits from the West, they did not foresee the depth and intensity of the U.S. response. This surprise can be taken as evidence of misperception by the Soviet Union. On the other hand, one cannot conclude that it would not have invaded had it forecast the U.S. reaction accurately. The decision to invade was almost certainly *not* contingent on an assessment of the likely Western response. Rather, the nature of regional politics apparently dictated the Soviet decision. Again, surprise can be taken as evidence that misperception affected expectations, but not that it affected choice.

Misperception and Contingency

Misperception can only affect the choice of an actor whose decision is contingent on the actions of others.[40] By crudely dichotomizing foreign policy choices into cooperation and defection, we can delineate two possible contingent strategies. An actor may be a "reciprocator," prepared to cooperate if others cooperate, and prepared to stand firm if others fail to bend. Alternatively, the actor may be an "opportunist," prepared to stand firm if others cooperate, but prepared to cooperate if they stand firm.[41] Each must assess the preferences of others because they are central to its own decision; its misperception may be one of two kinds: either that the other has a dominant strategy when it actually does not, or that the other has a contingent strategy when it actually does not.

When neither actor has a dominant strategy, there are four possible outcomes (see Figure 11): both actors reciprocate, neither actor reciprocates, or one reciprocates while the other does not, and vice versa. In a world of mutual reciprocity (a "tit-for-tat" world), each party is prepared to respond in kind to the actions it expects the other to take,

40. Steven J. Brams, "Deception in 2 × 2 Games," *Journal of Peace Science* 2 (Spring 1977): 171–203, analyzes the situations in which actors have an incentive to deceive others. One cannot simply call misperception the flip side of deception, however, and presume that situations in which the outcome is affected by deception constitute the universe of situations in which misperception affects the outcome. After all, misperception can occur even when the misperceived actor has no incentive to deceive. Moreover, this chapter is specifically concerned with assessing the implications of misperception for international conflict and cooperation, and the conclusions suggest that the situations in which actors have an incentive to deceive are not the ones in which misperception results in otherwise avoidable conflict.

41. Robert Axelrod dubs this strategy "tester." See Axelrod, *The Evolution of Cooperation* (New York: Basic Books, 1984).

Figure 11. The effects of misperception when neither actor has a dominant strategy

	The misperceived actor:	
	Opportunist	Reciprocator
The misperceiving actor: Opportunist has no dominant strategy, preferring: C when the other defects and D when the other cooperates	CHICKEN WORLD Misperception emboldens (leading to DC outcome) or frightens (leading to CD outcome)	NO-EQUILIBRIUM WORLD Misperception leads to either CC or DD outcome
Reciprocator has no dominant strategy, preferring: D when the other defects and C when the other cooperates	NO-EQUILIBRIUM WORLD Misperception leads to either CD or DC	TIT-FOR-TAT WORLD Misperception leads to either DD or its avoidance

cooperating when the other cooperates and responding to defection with defection. In a nonreciprocal world (one exemplified by the game of "chicken"), neither actor is prepared to respond in kind to the other (each will cooperate when it expects the other to defect and will defect when it believes the other will cooperate); reciprocal situations are similarly characterized by no-conflict or two-equilibria games. Situations in which one actor reciprocates and the other does not are *always* games without equilibrium outcomes.

Misperceiving a Dominant Strategy

The implications of an actor's erroneous belief that the other actor has a dominant strategy are detailed in Figure 11. In only two of these cases can misperception result in unnecessary war, in a DD outcome that would not have occurred had the actors accurately perceived one another's preferences. One of them occurs when the misperceiving actor is a reciprocator who mistakenly believes that the other has a dominant strategy of defection when, in fact, it too is prepared to reciprocate. It is exemplified by the vigilant status quo state that believes another status quo state to be an aggressor. In the second case of misperception that results in otherwise avoidable mutual defection, the misperceiver is an opportunist who believes the other actor to have a dominant strategy of cooperation. War results because the misperceiving actor cheats on an opponent believed certain to capitulate but actually willing to reciprocate. For instance, before World War I, Germany was perhaps an opportunist in assessing British preferences in July 1914. Germany might have been deterred had it seen that Britain was prepared to reciprocate, but Germany's misperception of Britain as a cooperator led to an otherwise avoidable war. Other examples of misperception—such as the North's invasion of South Korea in 1950 and the Soviet decision to place missiles in Cuba—typically fall into this category of an opportunist's misperceiving a reciprocator as a cooperator.

In these cases deterrence fails because the misperceiving opportunist incorrectly sees a reciprocator as a cooperator. The misperception causes an otherwise deterrable, and thus avoidable, conflict. But the previous deduction—that avoidable war is a result of a reciprocator's misperception of another reciprocator as a defector—suggests that deterrence can also fail because of *misperceived hardness*, and not just because of *misperceived softness*. In both cases the *misperceived* actor is a reciprocator. When the misperceived actor is an opportunist, misperception does not lead to the DD outcome.

When both actors have contingent strategies but one believes the other to have a dominant strategy, neither the misperceiving actor nor the misperceived one is likely to welcome the occurrence of a misperception that leads to mutual defection—not, that is, unless they most prefer to go to war. The misperceiving actor knows that its decision is contingent on what the other actor does. It thus needs to know what the other will do but has no reason to hide its own preferences. Similarly, the misperceived actor, prepared to respond in kind to both cooperation and defection, has no incentive to mask its true preferences. Such a reciprocator does not wish to be seen as a capitulator by an opportunist, and, if it does not want war, it does not wish to be seen as an aggressor by another reciprocator.

It is important to note that one actor's misperception of another as having a dominant strategy can also facilitate the avoidance of war and mutual defection. If an actor inappropriately believes that the other has a dominant strategy when it really does not, the misperceiving actor has a cue as to how it should behave. The misperception transforms a fluid situation in which there is either no equilibrium outcome or two equilibria into one with a single equilibrium; it provides a clear course of action for a misperceiver with a contingent strategy. It can facilitate cooperation—for example, when a reciprocator misperceives another reciprocator as a cooperator. The potential outcome of mutual defec-

Figure 12. Actor without a dominant strategy misperceives a dominant strategy

	Situation perceived by B Actor B			Actual situation Actor B	
	Cooperate	Defect		Cooperate	Defect
Cooperate	3, 3	1, 4	Cooperate	3, 3	2, 4**
Actor A			Actor A		
Defect*	4, 2**	2, 1	Defect	4, 2**	1, 1
	Called bluff			Chicken	

*Actor's dominant strategy
**Equilibrium outcome

tion is avoided, and the misperception facilitates the occurrence of the outcome of mutual cooperation.

The belief that an actor has a dominant strategy when it actually has a contingent one can also ensure the avoidance of war by leading the misperceiver to capitulate. When an opportunist wrongly perceives another nonreciprocator to be bent on defection, its misperception ensures the avoidance of mutual defection by frightening it into co-operating. In such situations the misperceived actor has some incentive to mask its true preferences in order to induce the other to capitulate. In a game of "chicken," for example, an actor has an incentive to mask its true preferences: to appear to have a dominant strategy of defection. In Figure 12, the actual game is "chicken," but actor B's misperception of the situation as "called bluff" leads it to cooperate because it believes A has a dominant strategy of defection. The misperception facilitates coordination since it leads B to cooperate and accept the equilibrium outcome in which it does not receive its best possible payoff. An ac-curate perception might have led B to expect the DC outcome and defect, or to signal its opponent that it would cooperate and thus ex-pect the CC payoff. Understanding the true nature of the situation would lead B to do everything possible to avoid the CD outcome, but misperception facilitates coordination and leads B to accept the equi-librium outcome in which it gets its third-best payoff. In other words, misperception makes B into an unknowing altruist. This situation is one in which actor A has an incentive to deceive B about its true pref-erences and so to induce B's charitable behavior.

The resolution of the Cuban missile crisis may be an example of such a situation. The Russians were certainly opportunists who cheated by placing missiles in Cuba on the assumption that the United States

Figure 13. Misperceiving an actor with a dominant strategy

The misperceived actor actually
has a dominant strategy of:

	Cooperation	Defection
The misperceiving actor: Opportunist has no dominant strategy, preferring: C when the other defects and D when the other cooperates	Outcome should be DC:	Outcome should be CD:
	Misperceiving actor sees no-equilibrium world or "chicken" world of two equilibria	
Reciprocator has no dominant strategy, preferring: D when the other defects and C when the other cooperates	Outdome should be CC:	Outcome should be DD:
	Misperceiving actor sees no-equilibrium world or "tit-for-tat" world	

would capitulate and accept their presence; the Soviets were also prepared to withdraw the missiles if they believed at any point that the
United States was prepared to launch an air strike. If we assume that
the United States was similarly opportunistic and not truly prepared
to go to war, what frightened the Soviets into backing down was their
misperception of the United States as bent on standing firm. Such a
misperception transformed a game of "chicken" into one of "called
bluff." Deception facilitates the avoidance of conflict rather than exacerbates the possibility of its occurrence. In this case the United States
had an incentive to deceive the Soviets, and the Soviet misperception
facilitated de-escalation and the avoidance of greater conflict.

Misperceiving a Contingent Strategy

The other major misperception for an actor without a dominant strategy is to assume that the other actor is also without a dominant strategy
(see Figure 13). Here, too, misperception can either lead to conflict or
facilitate cooperation. The situation is transformed from one in which

Figure 14. Actor without a dominant strategy misperceives other actor to have no dominant strategy

	Situation perceived by A Actor B				Actual situation Actor B	
	Cooperate	Defect			Cooperate	Defect*
Cooperate	3, 3	2, 4**		Cooperate	3, 3	2, 4**
Actor A				Actor A		
Defect	4, 2**	1, 1		Defect	4, 1	1, 2
	Chicken				Called bluff	

*Actor's dominant strategy
**Equilibrium outcome

the misperceiver has a clear course of action because it knows the other's dominant strategy into one where it is uncertain. The misperception is a necessary but not sufficient condition for causing conflict or facilitating cooperation. This occurs, for example, when the misperceived actor has a dominant strategy of defection and the misperceiving actor is a nonreciprocator. An accurate assessment by actor A in Figure 14 would lead to CD as the equilibrium outcome. Actor A's misperception, however, leads it to believe that DC is also an equilibrium outcome that can be achieved by A's convincing B that it will defect, thus forcing B to cooperate. Actor A's misperception can result in a DD outcome. In this case the misperception does indeed affect the decision of the misperceiving actor and exacerbates the inherent conflict. It is not, however, the sole cause of the otherwise avoidable mutual defection; rather, the outcome is a result of the combination of the misperception and the opportunist's obstinacy.

This analysis of misperception suggests a number of important conclusions. First, misperception can affect an actor's choice only when that actor's decision is contingent on the behavior of the other actor: misperception is irrelevant for an actor with a dominant strategy. Second, despite the fact that misperception can affect an actor's choice, it need not necessarily lead to an undesired war (a nonequilibrium DD outcome); it can also facilitate cooperation and prevent conflict. Moreover, misperception can cause conflict only if the *misperceived* actor has either a dominant strategy of defection or a contingent strategy of reciprocity. Third, in those cases in which misperception can cause conflict, the misperceived actor has no desire to mask its true preferences. When one actor does in fact wish to hide its true preferences, such a successful deception actually facilitates coordination and the avoidance of conflict.

Misperception and Sequential Choice

The foregoing analysis does not specify whether the actors choose simultaneously or sequentially. A criticism often made of game-theoretic work is that it presumes simultaneous choice and thus fails to capture the dynamic nature of international relations, which do not occur in a historical vacuum but in an ongoing context of action and reaction. Yet if international politics is indeed a dynamic sequence of actions and reactions, it becomes more difficult to argue that misperception is important in determining international outcomes. If misperception can only affect the decision of an actor whose choice is contingent on another's, one must add more assumptions to make a case for the importance of misperception in a world of sequential decisions. After all, when an actor with a contingent strategy is responding to another's actions, it already knows how the other has acted. An actor with a contingent choice simply takes its cue from the immediately preceding behavior of the other.

For misperception to matter in a world of sequential decisions, the misperceiving actor's choice must be contingent on the other actor's *future* behavior—on how the other will respond to its current choice. Moreover, one has also to add the assumption that the other's future choice will differ from its previous behavior. In other words, two assumptions must be added if one wants to argue that misperception matters in a world of sequential decisions: that the misperceiver's decision is contingent on the other's future choice, *and* that the other's future preferences will differ from those reflected in its past choices. Given these two additional assumptions, the previous analysis of the situations in which misperception matters can be generalized.

I argue above that a reciprocator's misperception of another as a cooperator leads to the avoidance of conflict. An example is perhaps provided by Neville Chamberlain's dealings with Germany in 1938. By the end of the Munich crisis, Chamberlain clearly knew Hitler's intentions regarding Czechoslovakia, but Chamberlain's decision about what to do was contingent on his expectations of Hitler's future behavior. Chamberlain was by then ready to reciprocate future German defection or cooperation but can be said to have been uncertain about Hitler's future preferences. His misperception was to believe Hitler's assurances that these were his last demands. Chamberlain did not extrapolate from

Hitler's past and current behavior; he made his contingent decision in 1938 based on his mistaken belief that Hitler would be a cooperator in the future.[42]

The argument about the importance of misperception in the Soviet decision to deploy missiles in Cuba must also include these additional assumptions if it is to place the Soviet decision in a broader context that includes action and reaction. The Soviets were opportunists whose decision was contingent on their calculation of the likely American response to their action. Their mistake was to extrapolate from American behavior during the Berlin wall crisis and thus expect the United States to cooperate and accept Soviet missiles in Cuba. In the Berlin crisis, however, the United States had had a dominant strategy of co-operation, whereas in the missile crisis it had a contingent strategy that led it to respond not by capitulating but by standing firm.

A change in the misperceived actor's preferences often lies at the root of the misuse of history. Incorrect historical lessons are often inappropriate extrapolations of others' preferences and behavior.[43] The historical interpretations may be, and indeed usually are, correct. A problem arises only when the misperceived actor's preferences change from what they have been in the past.[44] Even so, the misperception that results from such inappropriate extrapolation matters only when the misperceiver's decision is contingent on its expectations of the other's subsequent response.

Misperception and International Relations

The foregoing conclusions, derived from an analysis of all possible games, are obviously applicable to any subset that best exemplifies real-world situations.[45] In a major and wide-ranging empirical study of crisis

---

42. In fact, Hitler provided constant reassurance that he wanted peace and even pointed to his own racism to support his contention that he had no designs on territories not populated by Germans. The British problem was to ascertain if Hitler simply had racial motives for unifying Germans or had larger territorial ambitions.

43. Jervis, *Perception and Misperception in International Politics*, pp. 217–82; Ernest R. May, *"Lessons" of the Past: The Use and Misuse of History in American Foreign Policy* (New York: Oxford University Press, 1973); Richard E. Neustadt and Ernest R. May, *Thinking in Time: The Uses of History for Decision-makers* (New York: Free Press, 1986).

44. For an empirical assessment of how strategies change, see Russell J. Leng, "When Will They Ever Learn? Coercive Bargaining in Recurrent Crises," *Journal of Conflict Resolution* 27 (September 1983): 379–419.

45. Anatol Rapoport and Melvin J. Guyer, "A Taxonomy of 2 × 2 Games," *General Systems* 11 (1966): 203–14, enumerate the seventy-eight unique 2 × 2 games. Most do not have labels, and probably not all have real-world equivalents.

Figure 15. Nine crisis situations

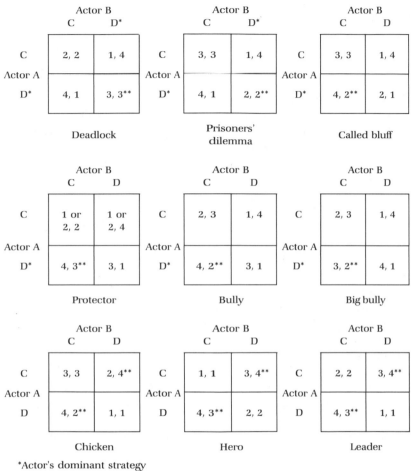

*Actor's dominant strategy
**Equilibrium outcome

dynamics, Snyder and Diesing find that all the historical events they analyze can be represented by nine games (Figure 15). Some of the nine are familiar ones that already have well-known labels: prisoners' dilemma, chicken, hero, and leader. The five other games they find to represent common real-world occurrences are those they call deadlock, called bluff, protector, bully, and big bully. They argue that some of these games (hero, leader, and protector) characterize the relationships of allies, and that others (bully, big bully, and deadlock) exemplify re-

lationships between adversaries. Three of the games (prisoners' dilemma, chicken, and called bluff) may characterize relationships
between either allies or enemies.

Snyder and Diesing provide numerous historical examples of misperception in both kinds of relationships. Their discussion of misperception is derived from their analysis of these historical cases, but it is
also possible to assess the impact of misperception through the formal
analysis of the nine games that, according to their argument, constitute
the universe of international crises. This analysis is accomplished by
considering the preference orderings of one actor (the row player, A)
for each of these games (see Figure 16).[46]

Actor A's preferences clearly define five of the nine situations in
which it need not bother to assess B's preferences—much less assess
them accurately—as long as it knows its own. When A knows, for example, that it prefers the outcome DC to CC to CD to DD ($T > R > S > P$
[T = temptation $> R$ = reward $> S$ = sucker $> P$ = punishment] for
those accustomed to a different terminology), the game must be
chicken. Moreover, if A knows its own preferences and knows that these
nine games constitute the universe of situations, it can sometimes deduce B's exact order of preference.

In the other four games, however, A's preferences alone do not define
the game. Thus A must accurately perceive B's preferences in order to
know what the game actually is. Otherwise, A might confuse bully and
deadlock, two games in which its preference orderings are the same.
Similarly, A might confuse prisoners' dilemma with called bluff. Even
in these four situations, misperception does not affect A's behavior or,
therefore, the game's outcome. In both bully and deadlock, A prefers
to defect regardless of B's actions. Mistaking one game for the other
will lead A to expect the wrong equilibrium outcome (DC in bully, DD
in deadlock) but will not change A's course of action. Since actor A also
has the same dominant strategy in both prisoners' dilemma and called
bluff, misperception again will cause A to expect the wrong outcome
but will not lead it to change its behavior.

Thus misperception cannot affect the behavior of the row player
(actor A) in any of the nine games found by Snyder and Diesing to

46. The analysis that follows assumes that actors know that these games constitute
the universe of crises. Although Snyder and Diesing's argument—that the universe of
crises they analyze reduces to these nine games—is convincing, there may, of course, be
other possible games that capture the essence of situations that occur in international
relations. Nevertheless, many preference orderings are nonsensical when applied to international politics; thus decision makers may implicitly understand that the universe of
possible situations is limited.

Figure 16. Preferences of actor A in nine crisis situations

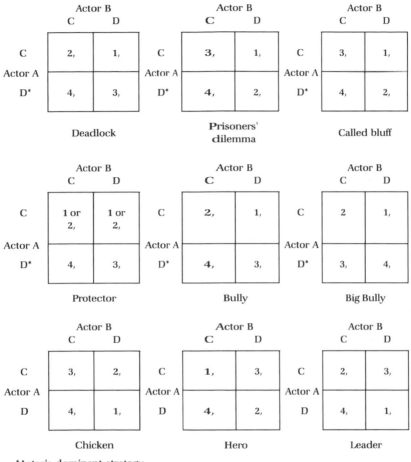

characterize international crises. There are five games that A cannot possibly mistake for one of the others. In the rest, misperception may lead A to mistake one game for another and thus expect the wrong outcome; but in each of these cases, A has a dominant strategy, and its course of action remains the same regardless of B's preferences.

For actor B, life is not quite as simple, since its preference orderings define just three of the nine games unambiguously (see Figure 17). Only in deadlock, prisoners' dilemma, and hero are B's preferences enough to let B know what game it is playing. On the other hand, actor B's

Figu

ERRATA

Why Nations Cooperate
by Arthur A. Stein

Figure 17 on page 80 was misprinted. The correct presentation is:

(

Actc

I

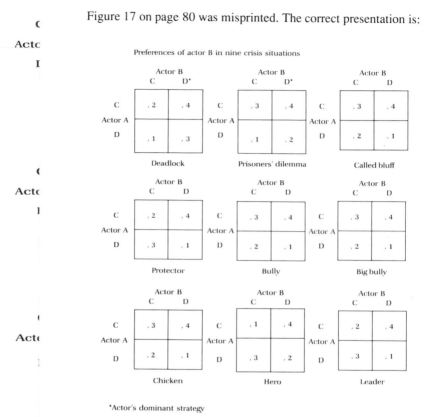

Preferences of actor B in nine crisis situations

*Actor's dominant strategy

(

Actc

]

(

Act(

]

*₁

preference orderings are identical in called bluff, bully, big bully, and chicken, and B must accurately perceive A's preference ordering to know what the situation is. Similarly, actor B's preference orderings are the same in protector as in leader, and here too a knowledge of A's preference ordering is necessary for B to know the true context.

Actor B's misperception of A's preferences, and its possible confusion of called bluff, bully, big bully, and chicken, poses a problem for it does not have a dominant strategy in any of these four games. Yet because actor A *does* have a dominant strategy (D) in three of the four (called bluff, bully, and big bully), B's behavior is not affected by its mistaking any of these three for one of the other two. B is forced to cooperate with A in these games, and the equilibrium outcome is always DC. B's error does not affect the game's solution.

If, however, B believes chicken to be one of the other three games, its misperception *may* affect its behavior and the game's outcome as well. If B believes the game to be called bluff or bully or big bully, it will automatically cooperate with A rather than stand firm (which would have produced its most preferred outcome by forcing A to back down). In other words, B makes the error of believing it is weaker because it mistakenly assumes A to have a dominant strategy.

Yet even when B's behavior is affected by its confusion of chicken with one of these other games, the outcome may not be affected. The outcome changes only when B happens to be the actor who would not have backed down first had both parties correctly understood the game to be chicken. There are two equilibrium solutions to the game of chicken, CD and DC, as well as the natural outcome, CC. Actor B's misperception guarantees the DC outcome, which is to actor A's advantage. The DD outcome is also a possible (albeit unlikely) outcome to the game of chicken, however; B's misperception, by forcing B to cooperate, thus ensures that they avoid the disastrous DD outcome.

Another possible misperception is B's belief that called bluff, bully, or big bully is actually chicken. Here the problem is the opposite of the one discussed above. Actor B does not recognize that A has a dominant strategy of defection that requires B to cooperate. B's accurate assessment of the situation would lead to a DC outcome, one of two possible equilibrium outcomes in the game of chicken (the other is CD). B's error affects the outcome only if B insists on defecting since it expects that A will cooperate. In this case CD would be the result. Actor A's dominant strategy, however, is to defect. Even if the misperception is sustained, the end result should be the same, for in chicken B must cooperate if A does not. In playing a game it believes to be chicken, B attempts to convince A that it will defect, while trying to determine whether A is more likely to cooperate or defect. This is a situation in which B should try particularly hard to determine A's true preferences. Even if the misperception continues, the search should at least suggest to B that A's commitment to defection is a strong one, thus inducing B's cooperation (result: DC). Thus, even if B is mistakenly playing chicken rather than one of the other three games, the outcome should be the same; B will concede as it becomes clear that A intends to defect. The misperception heightens tension in that the DD outcome will occur if B insists on defecting; but DC, the equilibrium outcome for the real game, typically emerges despite the misperception. Even if DD does occur, the misperception is only one part of the cause; it must be conjoined with B's obstinacy as well.

Finally, one other case of misperception can occur. Actor B's preference orderings are the same in both leader and protector. Once again, B has no dominant strategy, but the two games do differ with respect to A's position. Protector refers to a relationship between allies in which A is B's protector. In this situation A's dominant strategy is to defect, and B does not mind cooperating since it obtains its second-best outcome. Leader, also an alliance game, differs from protector in that A has no dominant strategy, and the two actors must coordinate their actions. If B mistakes a game of leader to be protector, B is shortchanging itself by allowing A to defect and reap the benefits. The misperception provides B with its second-best outcome rather than the more preferred outcome it might have been able to achieve. In effect, the misperception solves the problem of coordination by leading actor B to forgo the possibility of the CD outcome and accept the DC outcome (both are equilibrium solutions).

The reverse misperception is more problematic. If B perceives the game to be leader when it is really protector, B is in for a rude awakening if it attempts to lead by defecting. If B does defect, it hurts itself by ensuring its own worst payoff and A's second-best one. This misperception occurs among allies when the weaker party thinks of itself as being on an equal footing with the stronger one. The result is that the misperceiving actor deprives itself of the protection of the stronger party, a relationship of which it has the greater need.

The role of misperception in international relations is thus quite different from that suggested by Snyder and Diesing and many other misperception theorists. Misperception may well be a common occurrence in international relations and may often affect an actor's expectation of the probable outcome. But the assumption that misperception affects an actor's choice and thus changes a game's outcome does not always hold, for an actor's course of action depends only sometimes on its correct assessment of the other actor's preferences. In other words, misperception is often irrelevant to the cause and escalation of crisis and war. If the actors know that Snyder and Diesing's nine games exemplify the universe of crisis situations, there are instances in which neither actor requires an accurate knowledge of the other's preferences. For actor A, five of the nine situations are uniquely defined by a knowledge of its own preferences; for B, three of the games are so defined. Further, when A *does* mistake one game for another because it misas-

Figure 18. Mutual misperception

| | Actor B | |
	Dominant strategy	No dominant strategy
Dominant strategy	NO EFFECT A believes B does not B believes A does not	POTENTIAL DISASTER A believes B does B believes A does not
Actor A		
No dominant strategy	POTENTIAL DISASTER A believes B does not B believes A does	FACILITATES COORDINATION A believes B does B believes A does

sesses B's preferences, its error does not lead it to change its course of action. In each of the possibly confused games, A has a dominant strategy; assuming that actors will always act to maximize their own returns, misperception will not lead any with a dominant strategy to alter their behavior. Only their expectations will be affected.

Misperception changes a strict maximizer's decision only when its choice is contingent on the other actor's choice. This situation arises for the row actor, A, in none of these nine key games. For the column actor, B, on the other hand, this is a problem. Ironically, however, B's misassessment of A's preferences does not always turn crisis to conflict. B's errors exacerbate conflict only between allies. If B misperceives its adversary, however, the mistake will facilitate coordination between the two actors and thus allow them to avoid war.

Mutual Misperception

In addition to instances of a single actor's misperception of the other, it is possible for both actors to misread the other's preferences. There are three configurations of such mutual misperceptions that can occur (see Figure 18). If both actors have a dominant strategy but each believes that the other does not, misperception has no effect on their behavior or on the outcome, for each does what it would have done anyway. If neither has a dominant strategy but believes that the other does, their errors facilitate coordination. In this case each will try to feel the other out. If both insist on getting their own way, the DD outcome may result; if each mistakenly thinks the other has a dominant strategy, however, both are more likely to defer to the other, so that coordination will result.

The final class of mutual misperception is one in which only one of the actors has a dominant strategy, and each believes that the other's strategy is like its own. The actor with the dominant strategy believes that the actor without one also has a dominant plan of action. Because it does not realize that the other actor's plans are contingent on its own, it does not recognize the need to signal its true preferences. The actor without a dominant strategy, meanwhile, believes that the other actor also has contingent plans. This actor may try to get a better outcome by preempting the other player—behavior that seems to confirm the first player's misperception that the second actor also has a dominant strategy. This situation is ripe for disaster; here misperception may become the sole cause of an otherwise avoidable war.[47]

Conclusion

In the simplest terms, therefore, *misperception need not cause conflict* even when it does affect actors' choices and behavior.[48] Indeed, misperception can facilitate conflict avoidance, interactor coordination, and even mutual cooperation. For misperception to cause conflict, the misunderstood actor either must have a dominant strategy of defection or must be a tit-for-tat reciprocator. In either case it has no desire to mask its true preferences. If the misperceived actor recognizes the other's confusion, its incentive is to signal its true preferences. Moreover, the misperceived actor's only incentive to deceive will be its desire to facilitate the avoidance of conflict.

Implicit in discussions of the impact of misperception are assumptions about what guides the decisions of misperceiving actors; the study of the implications of misperception thus requires the explicit formu-

---

47. If states know their own preferences, they cannot misperceive a prisoners' dilemma to be chicken or vice versa. All too often, scholars suggest that states must be confused as to which game they are playing. But even if one misreads the other's preference ordering, the games are differentiable by one's own preferences as well.

48. The core conclusions derived in this chapter, and originally published in 1982, have been reaffirmed by more sophisticated modeling techniques developed in the past few years. The new wave in game theory, especially as employed by economists, has been the development of extensive-form games and especially games of imperfect and incomplete information. Like many other new techniques in the social sciences, these have been imported into political science and international relations. For examples of more formal work that reaches the same general conclusions, see Bruce Bueno de Mesquita, "The War Trap Revisited," *American Political Science Review* 79 (March 1985): 156–77; Robert Powell, "Nuclear Brinkmanship with Two-Sided Incomplete Information," *American Political Science Review* 82 (March 1988): 155–78, especially pp. 167–68; and James Morrow, "Capabilities, Uncertainty, and Resolve: A Limited Information Model of Crisis Bargaining," *American Journal of Political Science* 33 (November 1989): 941–72.

lation of these decision criteria. If the misperceiving actors are rational maximizers of their own returns, it is clear that misperception cannot affect the decision of an actor with a dominant strategy.[49] In such situations misperception can affect the actor's expectations, but not its actions or the outcome of the situation. Misperception affects only the decisions of an actor whose maximizing strategies are contingent on the actions of another. Even then, when misperception affects an actor's decision and is a determinant of the outcome, it need not necessarily result in conflict.

The conclusions about the consequences of misperception can also be illuminated by a simple thought experiment. Imagine a situation in which the preferences of actors are known to all and in which each makes a choice according to its own self-interest. These choices lead to a particular outcome. Then imagine that some of the actors either do not know or are mistaken about the preferences of others. Such misperception might lead, either by accident or because knowing others' preferences did not affect the actors' choices, to the same outcome as would have occurred in the case of full information. But in some cases such misperception would lead to deviations from the otherwise expected outcome.

The primary conclusion here, that misperception can lead to deviations from otherwise expected outcomes, has immense implications. The impact of misperception depends on the otherwise expected outcome. The liberal emphasis that misperception leads to otherwise avoidable conflict is accurate but presumes a world in which full knowledge would generate cooperation. But misperception can lead to otherwise avoidable cooperation when a world of full knowledge would generate conflict.

Hence, one cannot conclude that misperception causes conflict simply because it occurs in crises that result in war. Misperception may be coincidental to—rather than determinative of—the occurrence of war, because war can be an equilibrium outcome that results from specific configurations of actors' preferences. Even if misperception does sometimes play a causal role in the outbreak of war, its impact is situationally circumscribed. It is certain, therefore, that if one limits the empirical study of misperception to crises that do result in war, one ensures an inaccurate assessment of the overall impact of misperception in international relations.

---

49. Misperception can matter if actors focus on others' returns as well as their own. Subsequent chapters discuss other ways in which states assess their own interests.

The ambiguous impact of accurate perceptions on cooperation and conflict is true of all social interactions, not just of international politics. Although misunderstanding and misperception can cause otherwise avoidable conflict, full information does not guarantee cooperation and harmony. In fact, a certain amount of interpersonal ignorance may provide a lubricant of social interaction. Think what would happen if people could suddenly read each other's thoughts. William James, the psychologist and philosopher, believed "the first effect would be to dissolve all friendships." Or as columnist Jack Smith put it, "by nightfall human society would be in chaos."[50]

50. Jack Smith column, *Los Angeles Times*, January 22, 1986, pt. V, p. 1.

4

Extinction and National Survival

Survival is the nation's highest priority in a world that makes its extinction a possibility and that provides a historical record replete with examples of countries that have disappeared.[1] Realists, in particular, posit that a country's primary objective is to maintain its territorial and political integrity.[2] Conceptualizing states as entities formed to provide protection against external assault, they see defense and security as the most basic goods that governments provide.[3]

1. Indeed, scholars are quick to point out that the balance of power, often touted as the core of the study of international politics, provides no guarantees that states will survive. Ironically, the most recent incarnation of the realist argument holds neither that the balance of power ensures state survival nor that it ensures peace and stability, but merely that in a world in which self-interested states act minimally to guarantee their own survival, balances will recur. See Kenneth N. Waltz, *Theory of International Politics* (Reading, Mass.: Addison-Wesley, 1979).

2. Among others, see Stephen D. Krasner, *Defending the National Interest: Raw Materials Investments and U.S. Foreign Policy* (Princeton, N.J.: Princeton University Press, 1978), p. 41.

3. In the language of microeconomics, states are firms that produce protection. There is a long tradition of assuming the origins of the state to lie in the requisites of survival, with the ability to wage war as a central element. See Robert L. Carneiro, "A Theory of the Origin of the State," *Science*, August 21, 1970, pp. 733–38; Richard Bean, "War and the Birth of the Nation State," *Journal of Economic History* 33 (March 1973): 203–21; and Charles Tilly, "Reflections on the History of European State-making," in *The Formation of National States in Western Europe*, ed. Charles Tilly (Princeton, N.J.: Princeton University Press, 1975), pp. 3–83. Also see Elman R. Service, "Classical and Modern Theories of the Origins of Government," in *Origins of the State: The Anthropology of Political Evolution*, ed. Ronald Cohen and Elman R. Service (Philadelphia: Institute for the Study of Human Issues, 1978), pp. 21–34; Frederic C. Lane, "Economic Consequences of Organized Vio-

This concern with survival constitutes an important element of context and circumstance missing from most studies of strategic interaction, including the preceding chapters. Because these models assume that states' calculations involve assessing the payoffs associated with various strategies, they describe nations as choosing options divorced from context. Such analyses do not consider whether states fear for their own survival or anticipate the disappearance of their rivals. Nor does it matter whether states confront losses or gains. Rather, payoffs generate specific games which constitute situations with particular dynamics. How states evaluate and compare payoffs is assumed. No other information is required about the circumstances in which states choose between cooperation and conflict.[4]

But the possibility of disappearance can affect the decision calculus of both those whose existence is threatened and their rivals. States adopt policies of deterrence because they worry about extinction. Such fears may lead nations to act conservatively and not take risks for peace. Or a choice between unfavorable outcomes may instead lead them to take great chances; they may even act recklessly. States that expect to reap the rewards of another's departure, on the other hand, may choose conflict even when cooperation costs them less in the short term. Indeed, these contextual considerations about extinction are critical to states' determinations of whether to maximize their short- or long-term payoffs. In either case the results are situationally circumscribed; long-term thinking, for example, is no guarantee of present-day cooperation.

Conservative Rationality

States and individuals act conservatively and do not endanger their survival for the chance to realize small gains. Actors may choose not to maximize expected payoffs when confronting the possibility of exhausting their reserves. One implication of the potential for bankruptcy is that actors may not risk assets in gambles with positive expected outcomes. States concerned with ensuring their survival in an anarchic world may eschew policies with high expected returns that also threaten too high a risk of total loss.

---

lence," *Journal of Economic History* 18 (December 1958): 401–17; Roger D. Masters, "The Biological Nature of the State," *World Politics* 35 (January 1983): 161–93; and Ronald W. Batchelder and Herman Freudenberger, "On the Rational Origins of the Modern Centralized State," *Explorations in Economic History* 20 (January 1983): 1–13.

4. The relative positions of the states, their status quo points, and the consequences of the payoffs for them do not particularly matter. Moreover, the payoffs associated with outcomes are known with certainty rather than probabilistically.

One illustration of this phenomenon is provided by the following thought experiment. You are offered the opportunity to bet double or nothing one year's salary for a 50 percent chance to get twice that amount. In strict expected-utility terms, you should be indifferent between the certainty of getting your annual salary and a gamble for double it. Any additional increment, however, should lead you to prefer the gamble to the certainty of receiving your salary. Yet few individuals would take the gamble. The only ones who might are those with a reservoir of assets that make losing affordable. In short, in a choice between a sure gain and the possibility of nothing, actors take the sure thing.

The same situation holds for the person with five dollars who is given the opportunity to bet one dollar in a gamble that offers a ten percent chance of winning twenty-one. The expected payoff for the wager is one dollar and ten cents. Since the expected outcome exceeds the cost of the gamble, it would be rational to take the bet. But the person offered the wager has only five dollars. Is it still rational to risk a dollar? Would it be rational to bet a second dollar upon losing the first? The expected return of the gamble remains positive for every one of the five dollars. Yet there is some danger of losing all five before winning the payoff. Indeed, the odds of losing five times in a row are 59 percent. Despite the positive expected payoff for each of five successive wagers, there is an almost 60 percent chance of going bankrupt in this game.[5] It would be fully rational not to wager in such a case, therefore—to eschew an option with positive expected returns in order to avoid an outcome with too high a chance of bankruptcy.[6]

Such situations pose a dilemma. It is rational to gamble for gains and to maximize expected outcome. But it is also rational to avoid gambles that run the risk of ruin.[7] Being cautious with one's last dollar is hardly

5. Drawn from Robert P. Wolff, "Maximization of Expected Utility as a Criterion of Rationality in Military Strategy and Foreign Policy," *Social Theory and Practice* 1 (Spring 1970): 99–111.

6. The same point can be made about high wagers for small gains. Most people would not pay $498 for a lottery ticket with a one-half chance of winning $1,000, even though its expected value of $500 exceeds its cost. The classic explanation for this hesitancy is that actors are risk-averse because they do not value every dollar equally. Rather, they see each incremental dollar as worth less than the one before. Hence the incremental dollars that might be gained in such a gamble are worth less than the dollars that would have to be wagered to enter the lottery. An alternative explanation for such conservative rationality is not that actors are risk-averse but that they merely assess choices differently.

7. In a philosophical treatment of this argument, J. M. Blatt discusses the greedy but cautious criminal who fears the gallows; Blatt argues that a concern with survival poses a problem for expected utility theory; see Blatt, "Expected Utility Theory Does *Not* Apply

irrational. Risking the sovereignty and territorial integrity of the nation-state is not an action taken lightly.

Lexicographic Preferences

States obviously value their self-preservation, but the issue that arises is how they compare this objective with others. The decision to co-operate or conflict can entail outcomes that affect an array of state interests, and how actors compare different goals becomes central. Typically, they weight and aggregate them into a single metric. All they determine is some net expected value. But some options run high risks. As argued above, it is fully rational in such cases to pass up the option that maximizes expected payoffs in order to ensure survival instead.[8] This violation of the logic of expected outcome suggests an alternative logic, one in which survival is given preeminent weight and in which attractive gambles are undertaken only when more fundamental objectives are ensured. Such preferences are lexicographic ones. Actors with lexicographic preferences maximize in sequence rather than make trade-offs. They compare outcomes on the first objective and only then compare those that do equally well at that level with regard to how well they do on secondary objectives, and so forth. Thus they will choose an option that maximizes the main objective regardless of how it does on secondary ones; no option that fails to maximize a core objective will be chosen regardless of how well it ensures others.

Consequently, states may opt not to maximize expected outcome because they view goals hierarchically and evaluate them sequentially. States that place preeminent weight on security and do not gamble with it regardless of the temptation to do so may, for example, act to maximize assured security rather than expected payoffs. Such states would undertake attractive gambles only when assured of survival.

The "onion" theory of international objectives provides one example of a lexicographic view of state interests. Richard Rosecrance argues that there exists a hierarchy of state objectives, with security the most

---

to All Rational Men," in *Foundations of Utility and Risk Theory with Applications*, ed. Bernt P. Stigum and Fred Wenstop (Dordrecht, Holland: D. Reidel, 1983), pp. 107–16.

8. For a trenchant critique of expected utility theories of political participation, see James DeNardo, *Power in Numbers: The Political Strategy of Protest and Rebellion* (Princeton, N.J.: Princeton University Press, 1985), pp. 52–57. DeNardo favors the use of a general rational choice approach, in which different definitions of rationality are deemed appropriate as a function of the substantive application.

fundamental.[9] States will pursue secondary ideological objectives only when survival is ensured. Similarly, they will attempt to accomplish material objectives, which are tertiary, only when they have achieved their ideological goals.[10] Rosecrance describes the evolution of international politics as entailing the peeling (or growing) of the layers of an onion.[11] States that emphasize lexicographic preferences are rational, calculating, and even maximizing, but they do not maximize expected utility as classically defined.[12]

Framing

In addition to a conservative rationality that places preeminent weight on survival, there is a gambling rationality that accepts chances. Indeed, the same risk-averse actors who take sure things rather than wager may

9. Richard Rosecrance, *International Relations: Peace or War?* (New York: McGraw-Hill, 1973), chaps. 14 and 15. For a formal model using such a logic, see Charles W. Ostrom, Jr., "Balance of Power and the Maintenance of 'Balance': A Rational-Choice Model with Lexical Preferences," in *Mathematical Models in International Relations*, ed. Dina A. Zinnes and John V. Gillespie (New York: Praeger, 1976), pp. 318–32.

10. This formulation is much like Abraham Maslow's hierarchy of human needs, which posits that basic needs must be met before people pursue other ones.

11. Note that Rosecrance provides a deductive hierarchy of national interests. By contrast, Krasner, *Defending the National Interest*, pursues an inductive approach to assessing the national interest because he is dissatisfied with the inability of realism to delineate state objectives other than those of ensuring the sovereignty and territorial integrity of the state. Krasner argues that realist notions of the national interest are weak precisely because they do not establish secondary and tertiary interests that states pursue. Implicit in Krasner's argument, then, is the possibility that state interests are hierarchical and lexicographic.

12. The scholarly debate on the nature of peasant politics is essentially about the bases of peasant decisions and mirrors the argument here. The moral-economy school argues that peasants, concerned about their survival, are risk-averse actors who prefer sure things. In contrast, the political-economy school argues that peasants maximize expected payoffs and are willing to gamble. In this latter view, neither are peasants risk-averse nor do they decide lexicographically by maximizing security, subsistence, and survival. Unfortunately, the debate is sometimes framed as one between those who view peasants as rationally self-interested and those who see them as acting in terms of a particular moral order. My own view is that the two perspectives describe alternative ways in which actors assess their self-interest. One is neither more moral nor more rational than the other. In terms of the argument I develop here, scholars must establish the bases of calculation and assessment employed by actors in different circumstances. People are purposive and calculating but can find compelling reasons to maximize different things in different situations. For the debate about peasants, see James C. Scott, "Exploitation in Rural Class Relations: A Victim's Perspective," *Comparative Politics* 7 (July 1975): 489–532; James C. Scott, *The Moral Economy of the Peasant* (New Haven, Conn.: Yale University Press, 1976); and Samuel L. Popkin, *The Rational Peasant: The Political Economy of Rural Society in Vietnam* (Berkeley and Los Angeles: University of California Press, 1979).

also be risk-acceptant when they confront a choice between sure losses and a gamble that may allow them to avoid loss altogether. Cognitive psychologists refer to this phenomenon of viewing losses and gains differently as "framing."

An array of social-psychological experiments consistently finds that individuals are risk-averse when making choices among different gains, but that they are risk-acceptant when confronted by losses. Most people will take a sure gain of $3,000 rather than an 80 percent chance of winning $4,000, even though the expected payoff of the latter is greater ($3,200). On the other hand, when people confront a choice between a sure loss of $3,000 and an 80 percent chance of losing $4,000, they choose the gamble rather than the certain outcome. People seem to prefer risks to sure losses but sure gains to potential ones.[13]

The Japanese attack on Pearl Harbor provides an excellent example of risk-acceptance. In 1941 the Japanese confronted a deteriorating status quo. The United States had embargoed shipments of oil to Japan and was pressing Japan to give up its Asian conquests. The United States was unwilling to cede much to the Japanese. The Japanese thus confronted a sure loss that they could conceivably accept. Alternatively, they could take a risky gamble, as they had in the Russo-Japanese War, which they won. They decided to take another leap in the dark; as War Minister Tojo put it in 1941, "Sometimes a man has to jump from the veranda of Kiyomizu Temple, with his eyes closed, into the ravine below."[14] As they fully recognized, the United States would certainly win

13. Amos Tversky and Daniel Kahneman, "The Framing of Decisions and the Psychology of Choice," *Science*, January 30, 1981, pp. 453–58. For a review of the literature on this "preference reversal" phenomenon, see Paul Slovic and Sarah Lichtenstein, "Preference Reversals: A Broader Perspective," *American Economic Review* 73 (September 1983): 596–605. For an experimental demonstration that decision makers allocate more to defense when framing implies high deficits in security, see Roderick M. Kramer, "Windows of Vulnerability or Cognitive Illusions? Cognitive Processes and the Nuclear Arms Race," *Journal of Experimental Social Psychology* 25 (1989): 79–100.

14. Chihiro Hosoya, "Characteristics of the Foreign Policy Decision-making System in Japan," *World Politics* 26 (April 1974): 354–55. Hosoya's point about the success of a previous risky venture raises the issue of how policymakers learn from history. Amos Tversky and Daniel Kahneman have shown that individuals miscalculate the probabilities of events because they place too much weight on personal experience and specific events; see Tversky and Kahneman, "Judgment under Uncertainty: Heuristics and Biases," *Science*, September 27, 1974, pp. 1124–31, and Daniel Kahneman, Paul Slovic, and Amos Tversky, eds., *Judgment under Uncertainty: Heuristics and Biases* (New York: Cambridge University Press, 1982). Nonetheless, the fact that they had won their gamble by going to war with Russia did not blind the Japanese to the knowledge that they were again embarking on a very risky route. Nor did it lead them to overestimate their odds of winning.

a protracted war if it fully mobilized its military potential.[15] But there existed some chance that a quick massive Japanese victory would lead the Americans to negotiate on better terms than the United States was currently insisting on.[16]

The alternative that nation-states confront when faced by certain losses is sometimes posed as a choice between war now and war later. When states confront a deteriorating status quo and perceive a high likelihood that a future war will come on unfavorable terms that will offer only the opportunity for total capitulation, they will often choose to run very high risks by waging war immediately.[17] The choices of capitulation, war now, and war later in a context of deterioration will often lead to the choice of war now, as one interpretation of World War I suggests. The European nations all confronted a choice between capitulation (the certainty of a diplomatic defeat) and war, which held out the possibility of victory.[18] They chose the possibility of victory rather than the certainty of defeat.

The decision to risk war can be similar to the decision to wage it. During the Cuban missile crisis, President Kennedy escalated and risked nuclear war merely to remove Soviet missiles from Cuba. Accepting those missiles would have entailed a loss of American prestige and perhaps posed some political danger to his administration but would probably not have affected the two nations' actual military balance. Still, the president framed the decision as one between war now and war later. He believed that a failure to respond to their hostile actions would embolden them while diminishing America's global stature and thus lead ultimately to armed hostilities. Consequently, he preferred immediate resistance even at the risk of nuclear war.[19]

15. Nobutaka Ike, ed., *Japan's Decision for War: Records of the 1941 Policy Conferences* (Stanford: Stanford University Press, 1967).

16. This interpretation differs from that offered by Bruce M. Russett, who argues the Japanese decision can still be explained as one of maximizing expected outcome because they chose "the least unattractive course of action from a set of options"; see Russett, "Pearl Harbor: Deterrence Theory and Decision Theory," *Journal of Peace Research* 4 (1967): 89–105, quote from p. 99.

17. For a review essay linking decline and the risky decision to go to war, see Jack S. Levy, "Declining Power and the Preventive Motivation for War," *World Politics* 40 (October 1987): 82–107.

18. In essence, this argument is made by L. L. Farrar, Jr., "The Limits of Choice: July 1914 Reconsidered," *Journal of Conflict Resolution* 16 (March 1972): 1–23. Farrar conflates options and outcomes, however.

19. Jack L. Snyder interprets Kennedy's framing of the problem as one of war now versus war later as an inability to face trade-offs. He uses this conceptualization to support a cognitive interpretation of the Cuban missile crisis; see Snyder, "Rationality at the

Again, this suggests that calculating whether an attack on others carries positive or negative utility is a complicated exercise for states. They may not attack others when the expected payoff is positive because they prefer surer and smaller gains to the risks involved in striving for greater ones. States will not jeopardize their existence for small and uncertain gains. On the other hand, states may attack others when the expected outcome is negative because the low odds of success in war are preferable to a sure loss.[20] States may jeopardize core objectives and run great risks when the alternative is to accept certain losses.

Understanding that potential losses affect the willingness of states to make foreign policy decisions that entail greater risks also sheds light on the impact on choice of crises, events that combine surprise, short response time, and high threat.[21] The conventional wisdom holds that crises bring riskier and more inappropriate decisions. Yet the predisposition to make riskier choices when confronting losses suggests that the element of high threat alone can explain this phenomenon. Actors may choose to gamble even in situations that entail neither surprise nor the necessity of a short response time. In sum, risky decisions are not necessarily a function of crisis, for they do not require the conjoint occurrence of the factors that define a crisis. Rather, the decisions to take risks observed during crises are really a function of situations that entail potential losses.

Brink: The Role of Cognitive Processes in Failures of Deterrence," *World Politics* 30 (April 1978): 345–65. The argument presented here is that Kennedy did not avoid trade-offs but took the riskier course of action, despite the possibility that it carried a lower expected payoff, because he wanted to avoid a sure loss.

20. Contrast this view with that of Bruce Bueno de Mesquita, the most recent and insistent proponent of an expected utility view of all international politics. See Bueno de Mesquita, "An Expected Utility Theory of International Conflict," *American Political Science Review* 74 (December 1980): 917–31; *The War Trap* (New Haven, Conn.: Yale University Press, 1981); "The War Trap Revisited: A Revised Expected Utility Model," *American Political Science Review* 79 (March 1985): 156–77; and "The Contribution of Expected Utility Theory to the Study of International Conflict," *Journal of Interdisciplinary History* 18 (Spring 1988): 629–52. For a general overview, see Paul J. H. Schoemaker, "The Expected Utility Model: Its Variants, Purposes, Evidence and Limitations," *Journal of Economic Literature* 20 (June 1982): 529–63.

21. This classic definition is by Charles F. Hermann, "International Crisis as a Situational Variable," in *International Politics and Foreign Policy*, 2d ed., ed. James N. Rosenau (New York: Free Press, 1969), pp. 409–21. There are some slight definitional differences among crisis scholars; see Michael Brecher, "A Theoretical Approach to International Crisis Behavior," *Jerusalem Journal of International Relations* 3 (Winter–Spring 1978): 5–24; and his "State Behavior in International Crisis: A Model," *Journal of Conflict Resolution* 23 (September 1979): 446–80. For a review, see Ole R. Holsti, "Theories of Crisis Decision Making," in *Diplomacy: New Approaches in History, Theory, and Policy*, ed. Paul Gordon Lauren (New York: Free Press, 1979), pp. 99–136.

The analysis of any foreign policy decision (indeed, of any decision) must incorporate some knowledge of the status quo and of how the outcomes being compared will affect it. That is at the heart of the distinction between gains and losses.[22] It is not enough to know the rank order of the preferences, for their actual values also matter. Where actors stand and whether they stand to gain or lose both matter immensely.

Implications for Deterrence

States pursue deterrent policies to ensure their survival, threatening unacceptable retaliation in order to prevent others from taking certain steps. The logic of deterrence has been formally developed and has been extensively discussed. That there exists both a conservative and a gambling rationality when survival is at stake has profound implications for our understanding of the workings of deterrence.

Deterring another state in the domain of gains is relatively unproblematic. We expect states to act conservatively in the domain of gains, not to take risks for uncertain rewards. States that find the status quo acceptable are not likely to take chances for uncertain gains that may jeopardize what they might otherwise have.

This is equivalent to the observations, both formal and empirical, that ensuring an acceptable status quo for another state makes it easier to deter it. This squares with a central conclusion from formal analyses that suggest that deterrence is maintained when the value of waiting exceeds the value of striking another, and this relationship obtains when the value of peace (i.e., the value associated with not going to

22. Satisficing and incrementalism are path-dependent and therefore also context-dependent arguments of decision making. The conception of satisficing is due to Herbert A. Simon, "A Behavioral Model of Rational Choice," *Quarterly Journal of Economics* 69 (February 1955): 99–118. The conception of incrementalism comes from David Braybrooke and Charles E. Lindlblom, *A Strategy of Decision* (New York: Free Press, 1963). It can be shown, for example, that satisficing generates the same outcome as expected utility if one incorporates a term for the cost of information and decision making. (See William H. Riker and Peter C. Ordeshook, *An Introduction to Positive Political Theory* [Englewood Cliffs, N.J.: Prentice-Hall, 1973].) That is, people do not maximize because, rather than assess every option, they choose a satisfactory outcome at some point. They do this because searching through options is itself costly. Such a notion of bounded rationality differs fundamentally from the rational-actor model as embodied in maximization models. The assumption of maximization makes it possible to delineate an explanation that is independent of the process of decision making. The same maximal outcome is chosen regardless of the process. But models of bounded rationality that do away with the maximization assumption can lead to the selection of different choices as a function of the path taken in the course of evaluating alternatives.

war) exceeds the expected utility of striking first. Hence, even if the value of peace does not exceed the expected utility of striking, states will not strike, they will be deterred, as long as there is some positive value of peace and as long as there is uncertainty associated with the expected utility of striking. The recognition of this phenomenon has led some scholars to emphasize the importance of rewards and inducements in maintaining the peace. A few even suggest that appeasement can sometimes work. But trying to appease a potential aggressor with certain gains may not induce the desired behavior. Generating a conservative rationality in another state, getting it to accept sure things and eschew risk, requires that the prospect of certain gains be combined with the possibility of very costly failure. Appeasement does not induce conservatism unless credible risks attend aggression.

Just as it is easier to deter another state that exhibits a conservative bent in the domain of gains, it is harder to deter a nation that has a risk-acceptant rationality in the domain of losses. If states confront deteriorating circumstances and the prospect of absorbing certain losses, they will be willing to take gambles, even to start wars that they have only slim odds of winning. A military posture that generates a higher negative expected payoff may still not be enough to deter such a nation if by eschewing war it will absorb certain losses. In such circumstances states may still choose the high probability of losses in war to the certain loss that attends capitulation.

Put differently, in the domain of losses deterrence must be overwhelming. The combination of the losses and the odds must be so great in comparison to the certainty associated with capitulation that even risk-acceptant actors will recoil from adventurism. In such cases minimal deterrence may not be enough and the uncertainty of the fog of war may not deter. For risk-acceptant actors searching to avoid the costs of certain capitulation will take chances and start wars against long odds and even given the prospect of paying still greater costs when they lose than had they capitulated initially. Here deterrence must be overwhelming in order to succeed.[23] In short, the requisites of deterrence are given by the options confronting and the calculations of the state to be deterred.[24]

23. This is one way to understand the logic of the argument that you must ease for others the ability to back away so they do not lose face. What this does is minimize the cost of capitulation and make it as painless as possible. Rubbing it in only increases the costs of capitulation and thus increases the odds of a fight instigated to avoid that capitulation.

24. This argument is analogous to that which maintains it is important for the United States to pay attention to the strategic doctrine of the Soviet Union.

Survival and the Prisoners' Dilemma

The possibilities of bankruptcy and the differential logics that operate in the domains of gains and losses also affect the prospects for cooperation and conflict in the prisoners' dilemma. Conflicting rationalities already exist in the prisoners' dilemma, as discussed in earlier chapters. On the one hand, it is rational for each actor to adopt its dominant strategy, one that maximizes its returns regardless of the behavior of others. The outcome of mutual defection in the prisoners' dilemma is an equilibrium one, and unilaterally departing from a dominant strategy only makes the actor that does so worse off. Yet the game involves a dilemma, which arises because the equilibrium outcome that results from the mutual adoption of dominant strategies is Pareto-deficient. In other words, there exists an outcome that both actors prefer to the equilibrium one. Mutual cooperation is Pareto-superior to mutual defection. If rationality requires that actors maximize, then rationality requires achieving mutual cooperation.

There is therefore a dilemma, or, put differently, a conflict as to what to do. It is rational to choose one's dominant strategy. After all, a dominant strategy ensures the maximum possible outcome regardless of the other's choice. Yet it can also be rational to choose one's dominated strategy. After all, it is better to get one's second-best outcome than one's third-best outcome.[25]

The prisoners' dilemma arises from this conflict of rationalities. It is interesting precisely because rationality provides no clear and unambiguous guide to behavior. Actors interested in making comparisons and calculations to maximize their outcomes find themselves torn between two plausible and logical choices. Rational arguments can be made both for defection and for cooperation. It is a dilemma: a conflict between competing rationalities, between different ways of assessing interest and action.[26]

---

25. Note that it does not matter whether the cell entries are negative or positive. That is, it does not matter whether the status quo is preferred or whether losses or gains are being compared. All that defines the game (in its ordinal version) is the preference ordering. In the story from which the prisoners' dilemma game receives its title, the two prisoners confront unpalatable options, all of which make them worse off than they were before making the choice.

26. As befits the fact that both options, cooperation and defection, can be rationally chosen, psychological experiments invariably find that some people defect and others cooperate. See, for example, Anatol Rapoport and Albert M. Chammah, *Prisoner's Dilemma: A Study in Conflict and Cooperation* (Ann Arbor: University of Michigan Press, 1965). For an excellent and wide-ranging introduction and survey both to game theory and to the findings of experimental research, see Andrew M. Colman, *Game Theory*

A dynamic conceptualization of the prisoners' dilemma provides one purported noninstitutional resolution to this conflict of rationalities.[27] The prisoners' dilemma as presented thus far is static. The actors are presumed to interact once and to make their choices simultaneously. But most interactions are ongoing; they have a history and a future. Moreover, few situations involve simultaneous actions. Often, an actor decides on a course of behavior in response to another's actions. In a one-shot prisoners' dilemma, the actor choosing second should always defect. After all, if the other had cooperated in the immediately preceding move, defection guarantees the responder with the best rather than second-best outcome that would result from responding to cooperation with cooperation. On the other hand, if the first actor defects, the second must also defect if it is merely to suffer its second-worst rather than very worst result. But if the game continues beyond a single round, a strategy of conditional cooperation may be preferable.

Most international relationships also have a history and a future. They too occur in an ongoing context—despite the fact that individual leaders, governments, and nations come and go. There may not have been any record of a Gorbachev-Reagan relationship before the former's accession to power, but a relationship did exist between the United States and the Soviet Union. Anglo-French, Franco-American, and Franco-German relations (among others) have a longer history than that of any individual French republic (longer, indeed, than that of any French monarchy as well). Even when new nations come into being, as Israel did in 1948 after an almost two-millennia disappearance from the international scene, a historical background to Zionist diplomacy preceded the arrival of independence.[28]

Just as most international relations have a past, they may also have a future. International interactions are not like one-night stands that presume no expectations of a future meeting, much less of a future relationship. Individuals can run, change their identities, and hide, but nations cannot.

International interactions occur in an ongoing context, one that in-

and *Experimental Games: The Study of Strategic Interaction* (New York: Pergamon Press, 1982).

27. For institutional discussions, see Stephen D. Krasner, ed., *International Regimes* (Ithaca, N.Y.: Cornell University Press, 1983).

28. For discussions of Zionist diplomacy before the independence of Israel, see Alan R. Taylor, *Prelude to Israel: An Analysis of Zionist Diplomacy, 1897–1947* (New York: Philosophical Library, 1959); N. A. Rose, *The Gentile Zionists: A Study in Anglo-Zionist Diplomacy, 1929–1939* (London: Cass, 1973); and Yehuda Bauer, *From Diplomacy to Resistance: A History of Jewish Palestine, 1939–1945* (New York: Atheneum, 1973).

cludes information about which states cheat on one another and which keep promises. Unlike commercial exchanges, which can occur between parties that may never interact again, nations do have to continue dealing with one another. A merchant can scream at an obnoxious patron in full knowledge that this particular shopper might never return. But the business of any one client is rarely crucial, and the merchant can regain his or her composure before dealing with the next patron. Moreover, alienating one customer will not typically affect others. In international politics, however, expectations about future relations play a central role. The consequences of actions taken today will necessarily reverberate into the future. The very certainty that there will be continued interaction affects policymakers' assessments and can make a strategy of conditional cooperation rational.

Indeed, a number of scholars argue that incorporating future gains solves the prisoners' dilemma. They use iterated games, infinite games, and nonmyopic games in attempts to incorporate future returns into a current calculus.[29] Conditional cooperation emerges as rational if enough weight is attached to future returns.

If both actors prefer second-best to third-best in a single interaction, they will surely do so in repeated ones. Moreover, repetition can mitigate an actor's fear that the other will defect. After all, a cooperator can retaliate against a defector who, in a previous round, obtained its best outcome while the cooperator got its worst. Such cheating is possible only once.[30] Similarly, the potential cheater knows it can only reap

29. Michael Taylor, *Anarchy and Cooperation* (London: John Wiley, 1976); Robert Axelrod, *The Evolution of Cooperation* (New York: Basic Books, 1984); Steven J. Brams, *Superpower Games: Applying Game Theory to Superpower Conflict* (New Haven, Conn.: Yale University Press, 1985); Nigel Howard, *Paradoxes of Rationality: Theory of Metagames and Political Behavior* (Cambridge, Mass.: MIT Press, 1971); and Steven Smale, "The Prisoner's Dilemma and Dynamical Systems Associated to Non-Cooperative Games," *Econometrica* 48 (November 1980): 1617–34. A review of the experimental social psychology literature on the prisoners' dilemma also concludes that an important consideration for subjects is whether they view the interchange from a short-term or a long-term perspective. See Dean G. Pruitt and Melvin J. Kimmel, "Twenty Years of Experimental Gaming: Critique, Synthesis, and Suggestions for the Future," *Annual Review of Psychology* 28 (1977): 363–92. For a review of Axelrod focusing on its applicability to international relations, see Joanne Gowa, "Anarchy, Egoism, and Third Images: *The Evolution of Cooperation* and International Relations," *International Organization* 40 (Winter 1986): 167–86.
30. This assumes that one can tell the difference between cooperation and defection. A new question is whether one actor can ascertain whether the other's action represents cooperation or defection. See George W. Downs, David M. Rocke, and Randolph M. Siverson, "Arms Races and Cooperation," *World Politics* 38 (October 1985): 118–46; Jonathan Bendor, "In Good Times and Bad: Reciprocity in an Uncertain World," *American Journal of Political Science* 31 (August 1987): 531–58; Michael Balch, "On the Reciprocal Dynamics of an Arms Race: The Repeated Prisoner's Dilemma under Conditions of Imperfect Ob-

the benefits of defection once and must contrast its single-shot achieve-
ment of best over second-best with the benefits of sustained cooperation, of repeatedly obtaining its second-best outcome rather than its
third-best one.

Michael Taylor demonstrates that attaching a discount to future values and incorporating them into the current choice generates the conditions in which conditional cooperation (a strategy of tit-for-tat)
emerges as a preferable strategy (see Appendix 1). This ingenious argument transforms the comparison of defection versus cooperation as
a comparison of best to second-best, and of third-best to worst, into a
comparison of second-best to third-best. It does so by including future
payoffs that all feature this comparison. Only on the first round is best
being compared to second-best.

This formulation relies on the critical assumption that returns are
fixed and constant for all future interactions. Ironically, the strategy of
conditional cooperation is both future-oriented and fully myopic. The
superiority of tit-for-tat (TFT) is based on a sufficient concern with
future rewards. Yet tit-for-tat as a strategy is fully myopic; it entails no
forward-looking features. Rather, it stipulates no more than that each
actor does to the other what has just been done to it. It is mechanistic
and uncalculating.[31] Yet if payoffs vary over time, a strategy of conditional cooperation might not continue to be sensible.[32]

---

servability," presented at University of California, Los Angeles, May 27, 1986; and Jack
Hirshleifer and Juan Carlos Martinez Coll, "What Strategies Can Support the Evolutionary
Emergence of Cooperation?" *Journal of Conflict Resolution* 32 (June 1988): 367–98. See
also David M. Kreps, Paul Milgrom, John Roberts, and Robert Wilson, "Rational Cooperation in the Finitely Repeated Prisoners' Dilemma," *Journal of Economic Theory* 27 (August 1982): 245–52. Note that there is a difference between not knowing whether the other
actor has just defected and not knowing what strategy the other actor is using.

31. Tit-for-tat can easily be programmed into any servomechanism. Hirshleifer and
Coll argue that programming an organism to play tit-for-tat is still more complicated
than simply having it play cooperate or defect. They find that if any factor is included
as a cost of complexity which reduces the returns from a tit-for-tat strategy, then defection dominates conditional cooperation even in iterated games. See Hirshleifer and Coll,
"What Strategies Can Support the Evolutionary Emergence of Cooperation?"

32. If payoffs change from one time period to the next, it might make sense to look
forward and not just backward. Two actors that have both been playing the first game
for a long time by using a strategy of conditional cooperation might each decide to depart
from the strategy of conditional cooperation for just one game that had much higher
payoffs. Each would hope that the other would continue with TFT and that its own
defection would thus bring large returns. Alternatively, there might be situations in which
only one actor is tempted to cheat by the prospect of an unusually large gain. In this
case, myopia and farsightedness have different consequences. If the actor with the potentially large payoff remains myopic, cooperation is sustained through the tempting
phase. If the actor with the great temptation is farsighted and the other is myopic, the
exploitive outcome is possible and cooperation will break down. If both actors are far-

The second critical assumption is that interaction will continue.[33] This means, for example, that no actor will go bankrupt; in other words, neither need fear extinction nor expect the other to disappear. Thus an actor can afford to be taken advantage of once in the game.[34] But if the consequences of even once being taken advantage of are disastrous, expectation of future gains will not ensure its initial cooperation.[35]

Even in international relations, both the fear and the reality of national disappearance exist. Discontinuity, as well as continuity, mark world history. Empires and nations rise and decline, but they also appear and disappear.[36] And if one broadens one's definition of "disap-

---

sighted, cooperation may break down even if the tempted actor decides not to defect while the other defects to protect itself because of fear that it is about to be taken advantage of. This example makes clear, however, a theme of this book: lack of information and uncertainty can engender cooperation, and full information can often derail cooperation.

33. There is some debate as to whether or not it is necessary to assume the existence of an infinitely continuing game for a strategy of conditional cooperation to be a rational approach to playing the prisoners' dilemma. Those who believe that conditional cooperation requires an infinite game argue that it is rational to defect on the last move if the game has a fixed end. If it is rational to defect on the last move, it must be rational to defect on the penultimate move since there is no future action that can be affected by cooperating at that point. By backward induction, it becomes rational to defect on every move. Others argue that the game need not go on infinitely but that the end point must not be known precisely. For others, even when the end point is known, it is rational to cooperate now as long as the weight attached to future interactions is high enough. Such cooperation would collapse close to any known termination point, but it is difficult to say precisely when.

34. Hirshleifer and Coll show that tit-for-tat is not a "robustly successful strategy" if players compete in a contest of elimination; see Hirshleifer and Coll, "What Strategies Can Support the Evolutionary Emergence of Cooperation?"

35. Because the other actor is assumed to play TFT and cooperate on the first move, the equation in Appendix 1 that stipulates the weight that must be attached to future returns is a function of three variables, the three highest of the four cell entries in the prisoners' dilemma game. The size of the lowest payoff does not matter at all. It does not matter how bad that payoff is. Indeed, any comparison of strategies in an iterated framework includes only an assessment of three of the variables in the game. In an earlier work, Axelrod derived a measure of the conflict of interest in a prisoners' dilemma game, an index that is a function of all four cell entries in the game. Conflict of interest increases as the sucker's payoff (S) gets worse. But in the iterated formulation, only three cell entries are included in the calculus, and when the other actor is assumed to be a conditional cooperator, the sucker's payoff plays no part in the calculus. In short, the degree of conflict of interest can increase and the prospects for cooperation be unaffected. For Axelrod's work on conflict of interest, see Robert Axelrod, *Conflict of Interest: A Theory of Divergent Goals with Applications to Politics* (Chicago: Markham, 1970). For discussions of the impact of worsening the sucker's payoff on cooperation, see Robert Jervis, "Cooperation under the Security Dilemma," *World Politics* 30 (January 1978): 167–214; Chapter 2 above; Charles Lipson, "International Cooperation in Economic and Security Affairs," *World Politics* 37 (October 1984): 1–23; and Roger B. Parks, "What if 'Fools Die'? A Comment on Axelrod," *American Political Science Review* 79 (December 1985): 1173–74.

36. See J. David Singer and Melvin Small, *The Wages of War, 1816–1965: A Statistical*

pearance" to include not just extinction but departure from the status of major power as well, the number of the fearful expands.[37]

The strategy of conditional cooperation is undermined not only because some nations fear for their survival but also because other states look forward to driving them out of the game of international politics. The anticipation of future payoffs can include those that accrue when others are bankrupted or destroyed. The derivation of the superiority of conditional cooperation applies only to situations in which actors cannot disappear, however. Axelrod, for example, bases his game simulations on a specific payoff matrix in which none of the cell entries is negative (T = 5, R = 3, P = 1, S = 0). Suppose, however, that the outcome of mutual defection, the punishment outcome, is negative. Suppose, also, that each actor has some stock of assets or wealth. Finally, assume that the bankruptcy of one actor provides the other actor with its best outcome. In such a case a high weight attached to future monopoly profits can lead this second actor to defect in order to bankrupt the other (see Appendix 2). Similarly, firms have differential abilities to withstand the losses that can arise when competition generates prices below costs and a single company can obtain monopoly profits by driving its competitors out of a market. To ensure long-run competition, therefore, nations and international regimes widely prohibit short-term predatory practices.[38] In cases when others can be profitably bankrupted, a concern with future payoffs makes conflict rather than cooperation more likely in the short term.[39]

Handbook (New York: John Wiley, 1972), pp. 24–30, for a compilation of major powers and members of the interstate system.

37. What constitutes fear for survival can vary. States and people can be entirely extinguished. The word "genocide" has been coined for a reason. At times, however, states and people are fearful not of their physical extinction but for the maintenance of their culture, a way of life, a set of autonomous institutions. Hence states resist outsiders because of fear of political domination even when they do not risk physical destruction or economic ruin. On the other hand, there are also historical cases of states voluntarily merging with one another in a federal arrangement in order to ensure their survival. Much the same is true of firms. Corporations resist hostile takeovers but also merge voluntarily. Such institutional restructuring increases efficiency and profitability, both of which are presumably universal objectives of firms. Some takeovers are resisted not because of fear of unemployment and not because of an alternative assessment of maximum profitability, but because of a desire to retain a corporate identity and way of life.

38. The international equivalent is the ban on dumping.

39. Similar results would be obtained if the payoffs were not the same for both actors. In such a circumstance, one actor might be tempted to bankrupt the other even if both begin with the same levels of assets. This modification of the game entails eliminating the symmetry that is at the heart of Axelrod's conclusions. One way to do so is not to make the cell entries the same for the two players. The other way is to keep the cell entries the same but assume that the two players have differential initial assets, which

The requirements for sustaining conditional cooperation when bankruptcy is possible are more stringent, for tit-for-tat is preferred to defection only when the discount rate is high without being too high. The actor must value returns in the intermediate future sufficiently to rule out defection in the short term, but not value the very long term so highly as to defect in the short and intermediate term in order to drive the other out of the game.[40]

Thus the incorporation of future rewards into a current calculus fails to provide a guarantee of cooperation now. The fragile cannot fear extinction; for the strong, future rewards cannot be those that come from predation.

The Chain Store Paradox

States, like firms, are interested in excluding players from the game as well as in bankrupting those already playing. Great powers try to maintain barriers to entry into the club. Nuclear powers, for example, have worked to prevent nuclear proliferation. Further, states sometimes act to quash nascent drives for nationhood. Israel, for example, steadfastly refuses to tolerate the creation of a Palestinian nation.

Reinhard Selten's "chain store paradox" illustrates the conflicting demands of self-interest in a competitive context in which keeping others out of the game plays a critical role.[41] A chain store has many branches, each of which can be challenged. In any given period a single potential challenger must decide whether to mount such a challenge, and the chain store must then decide whether to undercut the new entry. There are a fixed number of periods, and in each one the potential rival knows the results of past decisions. The matrix of payoffs is given in Figure 19.

---

also extends the analysis by incorporating some information about the status quo that precedes the game.

40. This outcome has been obtained by transforming the comparison back to one of best versus second-best outcome rather than second-best versus third-best. Such manipulations can be performed endlessly. Cooperation will be preferred following any transformation of the choice between cooperation and defection into one between second-best and third-best outcomes. If the choice is framed as being between best and second-best, or between third-best and worst, then defection will be preferred. For any sequence of interactions that involves some combination, there will be a critical threshold for the variable that determines how much weight is given to each outcome relative to the other.

41. Reinhard Selten, "The Chain Store Paradox," *Theory and Decision* 9 (1978): 127–59. For an application of the chain store argument in international politics, see James E. Alt, Randall L. Calvert, and Brian D. Humes, "Reputation and Hegemonic Stability: A Game-Theoretic Analysis," *American Political Science Review* 82 (June 1988): 445–66.

Figure 19. The chain store paradox

| | | Potential entrant | |
		In	Out
Chain store	Cooperative	6, 6	14, 3
	Aggressive	0, 0	14, 3

The potential rival must decide whether to enter the market currently served by the chain store. If it stays out, the chain store reaps its highest return and need make no decision. The potential rival obtains whatever returns are available for alternative uses of its capital. If the rival decides to enter the fray, however, both its returns and those of the chain store are determined by the latter's response. If the chain store cuts its prices to drive the challenger out of the market, each reaps returns of zero, the worst outcome for both. If, on the other hand, the chain store does not respond aggressively, the rival makes more money than it would through alternative investments and the chain store makes more than it would by slashing prices, but less than it would had there been no market challenge.

The decision of the potential challenger is fully contingent on its forecast of the chain store's likely response. If it expects the chain store to respond aggressively, it should pack its tent and go elsewhere. If it believes the chain store will respond cooperatively, however, it should go ahead and enter the market. The chain store, on the other hand, prefers that the challenger not enter. It has every incentive to threaten reprisal to deter a rival's entry. But after the other has already decided to enter, the chain store hurts only itself by going through with its threat and responding aggressively. Indeed, one reason for punishing the entrant is to signal still other potential rivals that it will deny entrants the fruits of the market. The desire to maintain monopoly rents in other districts or in future time periods leads the chain store to respond aggressively and to forgo some profits.

The classic game-theoretic solution is that in each time period the challenger does indeed enter the market and the chain store does cooperate. This result is obtained by backward induction. On the last move, the chain store would have no incentive to respond aggressively since it would hurt only itself. Knowing this, the challenger would obviously enter the market. On the next-to-last move, the chain store also has no incentive to respond aggressively. After all, an aggressive response to entry is intended to deter future entrants, and since the results of the last round are already known, there is no point in attempting deterrence. So the chain store will cooperate on the next-to-last move, and the potential rival will enter the market on the next-to-last move. Continuing this line of reasoning by backward induction, one concludes that the chain store will never respond aggressively and a rival will enter some market on each move. If there are 20 rounds, the potential challenger in each round will decide to enter and will obtain a payoff of 6. On each round the chain store will respond cooperatively and over the 20 moves it will earn a payoff of 120 (20 rounds times 6 for each round).

Yet as Selten points out, such a result is absurd. It is more likely that entrants will fear an aggressive response in the early rounds and not enter. If any do, the chain store is likely to respond aggressively to deter future entrants. In the specific example, each time the chain store punishes an entrant, it forgoes a payoff of 6. Yet each time it successfully deters an entrant, it obtains a payoff of 14. In this particular example the payoffs of each deterrence round justify two aggressive responses. If by punishing one or two early entrants the chain store can successfully deter entry in 10 of the first 12 rounds, it will earn a payoff of 188 (10 rounds of 14, and 8 rounds of 6). Selten dubs this the deterrence argument, one he finds just as rational as the induction argument. In fact, he states that most people share his view that the deterrence argument is more convincing and plausible, and that even "mathematically trained persons recognize the logical validity of the induction argument, but they refuse to accept it as a guide to practical behavior."[42]

At the heart of Selten's deterrence argument is the view that the ability to obtain higher future returns becomes the basis for accepting less in the short term. The desire for monopolistic rents becomes the basis for short-term self-abnegation. Actors make and, more important, carry out threats that carry short-term costs because of the long-run payoffs of keeping others out.[43]

42. Ibid., p. 133.
43. Economists have recently focused on the importance of reputation and the rational bases of predation; see, for example, David M. Kreps and Robert Wilson, "Reputation and Imperfect Information," *Journal of Economic Theory* 27 (August 1982): 253–79, and Paul Milgrom and John Roberts, "Predation, Reputation, and Entry Deterrence," *Journal of Economic Theory* 27 (August 1982): 280–312. For reviews of this burgeoning literature, see Robert Wilson, "Reputations in Games and Markets," in *Game-Theoretic Models of Bargaining*, ed. Alvin E. Roth (New York: Cambridge University Press, 1985), pp. 27–62; Louis

The conclusion to be drawn from the chain store paradox is very much like the previous discussion of the iterated prisoners' dilemma. The fear of extinction and the prospects of driving others out of the game become the stimuli for defection rather than conditional cooperation in the iterated prisoners' dilemma. In the chain store paradox the prospect of keeping others out of the market plays the same role.

In sum, deterrence has a temporal dimension. If a state's decision to cooperate or conflict at any point in time can depend on its assessment of the future and the weight it attaches to future payoffs, the ability to deter depends critically on the temporal horizons of those to be deterred. Inducing cooperation requires that rivals not envision each other's disappearance.

Short- versus Long-Term Calculations

The foregoing discussion of extinction implicitly recognizes that differing time horizons can provide conflicting bases for assessing self-interest. People do not always make calculations about the future using similar notions about it. Indeed, the conflict between present and future is ubiquitous. People differentiate between short-term and long-term self-interest. They defer gratification. In fact, a major issue for economists is to explain why some people consume immediately but others save and invest for future returns. Similarly, sociologists refer to differences in the propensity to defer gratification in order to explain why some groups lift themselves out of poverty and others do not.[44] We tell people not to be shortsighted, to think about the long haul. Actors may choose different strategies depending on whether they are trying to maximize short-term or long-term self-interest. The village idiot who always chose a nickel when offered that or a dime was actually quite rational. Townsfolk ridiculed the man as a fool and enjoyed displaying his behavior to newcomers. Finally, someone asked him why he always picked the nickel. "If I took the dime," he answered, "they would never offer me the choice again."[45] In cases such as this, the

Phlips, *The Economics of Imperfect Information* (New York: Cambridge University Press, 1988), chap. 7; John Roberts, "Battles for Market Share: Incomplete Information, Aggressive Strategic Pricing, and Competitive Dynamics," in *Advances in Economic Theory: Fifth World Congress*, ed. Truman F. Bewley (Cambridge: Cambridge University Press, 1987), pp. 157–95; and Jean Tirole, *The Theory of Industrial Organization* (Cambridge, Mass.: MIT Press, 1988), chaps. 8 and 9.

44. Shlomo Maital, *Minds, Markets, and Money* (New York: Basic Books, 1982).

45. Story told by Abraham Kaplan in his lecture "Irrationality in Decisionmaking," presented to the Jacob Marschak Interdisciplinary Colloquium on Mathematics in the Behavioral Sciences, University of California, Los Angeles, May 16, 1986.

criterion of maximizing self-interest is inadequate to determine which course of action will be chosen.

Different time horizons represent a problem not merely for individuals who must make decisions but for two or more interacting with one another. If actors have different time horizons when they interact, their varying perspectives can become the bases for conflicting assessments of self-interest and so generate dilemmas of cooperation and conflict. Different temporal views can exacerbate either cooperation or conflict. Alexandroff and Rosecrance maintain that deterrence in Europe in 1939 failed for precisely this reason. They argue that Britain shifted from a policy of appeasement to one of deterrence when, on March 31, 1939, it extended a guarantee of aid to Poland in the event of aggression. Because Britain had no effective sanctions to employ in the short run, it approached deterring German aggression as a long-term problem. Able to do little immediately to help nations on the continent facing Hitler's military, Britain committed itself to a long-term military buildup which should have deterred German aggression. But Germany took a short-term view of the situation and launched a strike on Poland in 1939. So Alexandroff and Rosecrance conclude, "if opponents are to attain 'mutual deterrence,' they must operate on roughly the same time horizon. Long-term maximizers will not always deter short-term maximizers, and vice versa. Britain was not deterred from giving the guarantee by her short-term weakness in Eastern Europe. She hoped to prevail in a long war in which Germany would take the offensive. Germany was not deterred from attacking Poland by long-term uncertainties; rather, she focused on short-term strength."[46] In effect, they argue that Britain would have acquiesced to German designs on Poland if both Britain and Germany had been taking a short-term view, and that if both had taken a long-term view, Germany would have halted its challenge when Britain signaled "no more" in the Polish guarantee of March 1939.

Much the same argument can be made about the otherwise inexplicable Japanese attack on Pearl Harbor. After all, the Japanese recognized that they could not win a protracted war with the United States. In other words, the United States possessed a long-term deterrent against a Japanese attack. But it did not possess a short-term one. The Japanese decided to gamble that inflicting a major blow on the United States would convince the Americans not to wage the long war

46. Alan Alexandroff and Richard Rosecrance, "Deterrence in 1939," *World Politics* 29 (April 1977): 404–24, quote from p. 421.

that it could certainly win.[47] Any argument that war occurs when the
expected outcome is positive is incomplete, therefore, because it includes no assessment of the temporal horizon on which such an assessment is made.

Misperception and Extinction

The potential for national disappearance reinforces the twin arguments
derived in Chapter 3 that misperception can lead not only to otherwise
avoidable conflict but to cooperation as well. On the one hand, conflict
is exacerbated when a state assumes that time or history is on its side.
A state that wrongly believes that it can drive another actor out of the
game may defect and await monopolistic returns. So, too, the belief that
others' time on the planet is limited may lead states to await the demise
of rivals and opponents. Arab states will not recognize Israel and nor-
malize relations as long as they envision a future without it. Conflict is
also exacerbated when a state fears for its survival: when it fears being
driven out of the game by the exploitation of others. Here Israel provides
an example of a state unprepared to take risks for peace. Conflict in
the Middle East is sustained, and the prospects for peace diminished,
therefore, by the interaction between one state that fears extinction and
a group of others looking forward to that nation's demise.

If there exists no possibility that Israel will disappear, the misperception that it might causes otherwise avoidable conflict. If so, only the
perceptions, and not the payoffs, need change for a different dynamic
to develop. A strong Israel, unworried about its survival, will be able to
risk cooperation without fear of exploitation. Similarly, cooperation will
become possible if, with the passage of time, the fewer and fewer adults
who remember an all-Arab Middle East come to accept the reality of
Israel and do not look forward to its extinction.

Misperception can also cause otherwise avoidable cooperation, however. States may misassess the permanence of themselves and of others.
Such misperceptions lead them to cooperate when an accurate perception would lead to defection. Ironically, international cooperation
may be sustained by the mistaken belief that all the states that now
exist will continue to do so forever.

47. Russett, "Pearl Harbor." See also Chihiro Hosoya, "Miscalculations in Deterrent
Policy: Japanese-U.S. Relations, 1938–1941," *Journal of Peace Research* 5 (1968):
97–115.

Conclusion: Survival, Disappearance, and Dilemmas

The possibility of extinction generates dilemmas both for states interested in survival and for predators. When they interact, each can have compelling logics for both cooperation and conflict. Rationality provides neither an unambiguous prescription for, nor explanation of, behavior. The fear of bankruptcy and the temptation to cause another's extinction transform the incentives that would otherwise exist for cooperation and for conflict. That states can disappear, or can change from being major players to insignificant ones, means that international relations contains situations in which states depart from what our expectations would otherwise be about their rational behavior.

States that fear for their survival may rationally depart from the otherwise rational. Not wishing to risk their very national existence, they may be conservatively rational, eschewing gambles with higher payoffs. This is one way to view the unwillingness to cooperate in the prisoners' dilemma. Even when a state has reason to believe that the other will reciprocate, it may still make sense for it to defect and not run the great risk of being suckered. States may not take risks for peace if their national survival is at stake. Threatened states may also opt for a gambler's rationality and be willing to accept great risks with worse expected payoffs than those attending the sure loss they wish to avoid. The choice between cooperation and defection, then, is especially problematic for states facing the possibility of extinction and for those confronting a choice of losses that may or may not imperil national survival. It is rational to cooperate to achieve higher returns or ensure lesser rather than greater certain losses. Yet it is also rational to defect, to take the long shot that one can avoid losses entirely, in order to avoid the disastrous sucker's payoff.[48]

The possibility of bankruptcy also poses a dilemma for the potential predator. It is rational to cooperate and obtain higher returns. But it is also rational to defect and await the still-higher returns that will accrue when others are bankrupted or kept from entering the game.

Increases in national insecurity and uncertainty thus affect the pros-

48. The formulation in this chapter represents a richer view of the "security dilemma" confronting states, which is traditionally defined as a situation in which actions taken by one state to increase its own security trigger reactions by others that leave it worse off. In the parlance of this chapter, this classic security dilemma is between a short-term and a long-term basis for calculation. But this chapter delineates a number of security dilemmas, situations in which there are competing logics for national strategy. In such cases, choosing to ensure security in one sense leaves one worse off in another.

pects for international cooperation and conflict. Global depressions in which states confront grim choices are likely to be periods in which states opt for the gambler's rationality. Periods marked by the superiority of offensive weapons are ones in which states do not take risks for peace.[49]

States that think about the long run do not necessarily cooperate with others. Although they will not defect from fear of exploitation and extinction, they will do so if their view of the long term is one in which their rivals have been buried and destroyed. When actors have short time horizons, they are less likely to cooperate in the prisoners' dilemma. Yet long time horizons may be conducive to cooperation only in an environment in which it is unlikely that one actor can drive another out of the game and reap monopoly benefits, as it were. On the other hand, if an actor can envision driving another out of the game, as for example, in an environment in which the returns to competition are negative, then farsightedness can increase the chances of defection rather than cooperation.

There is an asymmetry in the burden of ensuring peace and cooperation in the international system. States concerned about their survival cannot be expected to take risks for peace. States that can exploit others must work to assure them that they will not do so.

The requisites of deterrence also differ. It is more difficult to deter aggression intended to avoid loss than it is to deter aggression pursued for gain. Ironically, it takes less to deter adventurous aggressors than it does the fearful and desperate. Further, deterrence must operate at the temporal horizon of the actor to be deterred.

Hence foreign policy analysis and international relations theory must incorporate more than the structure of the payoff environment. The payoffs associated with outcomes, the constellation of preferences and whether they represent gains, losses, or both, clearly affect the choices that states make. But the context in which decisions are made— whether their situations have been deteriorating and the kinds of futures they can look forward to—also matters immensely. Cooperation and conflict are the products not just of the payoffs but of the ways in which those payoffs are assessed.

The realist emphasis on survival is well placed. Nations concerned with their own extinction may indeed eschew otherwise potentially

---

49. Jervis triggered a literature on this subject with his "Cooperation under the Security Dilemma." My argument here is that the technological underpinnings may not change the preference ordering, but that the difference between absolute gains and absolute losses is the basis for different propensities for cooperation and conflict.

rewarding cooperation. But so, too, will predator states. The fear of weak, threatened countries conjoins with the greed and temptation of strong, secure ones to create a more conflictual world than might otherwise exist. This prospect of predation guarantees that a simple liberal emphasis on the long term is insufficient to ensure cooperation.

Self-interested states, concerned minimally with maintaining their physical and territorial integrity in an anarchic environment, confront a dilemma. The problem is not so much that they must rely on themselves but that their survival is not ensured. The international system provides no safety nets or insurance policies.

Appendix 1
Shadow of the Future

Value of defection in an iterated prisoners' dilemma given a strategy of conditional cooperation by the other actor (obtaining T on the first move, and P thereafter, and attaching a discount to future returns):

$$V(D/TFT) = T + wP + w^2P + w^3P \ldots$$

$$= T + wP/(1 - w)$$

Value of conditional cooperation given a strategy of conditional cooperation by the other actor (obtaining R on every move, but discounting future returns):

$$V(TFT/TFT) = R + wR + w^2R \ldots$$

$$= R/(1 - w)$$

The expected value of conditional cooperation exceeds that of defection if:

$$w > (T - R)/(T - P)$$

where T = temptation > R = reward > P = punishment > S = sucker; w = weight attached to future payoffs

NOTE: Michael Taylor, *Anarchy and Cooperation* (London: John Wiley, 1976), p. 89. The notation used here, as well as the phrase "shadow of the future," comes from Robert Axelrod, *The Evolution of Cooperation* (New York: Basic Books, 1984), p. 208.

Appendix 2
Iterated Prisoners' Dilemma with Bankruptcy

In this case, one actor sees the choices as follows:

$$V(TFT/TFT) = R + wR + w^2R \ldots$$

$$= R/(1-w)$$

$$V(D/TFT) = T + wP + w^2P \ldots + w^kP + w^{k+1}T + w^{k+2}T \ldots$$

$$= T + wP/(1 - w) - w^{k+1}P/(1 - w) + w^{k+1}T/(1 - w)$$

where k is the number of periods it takes to drive the other out of the
game, and thus the value of defection is T on the first move, P on the
moves until the other is bankrupted, and then T on all moves after
the other's departure from the game.

Now, defection is preferred to cooperation when:

$$R/(1 - w) < T + wP/(1 - w) - w^{k+1}P/(1 - w) + w^{k+1}T/(1 - w)$$

Multiplying through by $(1 - w)$, and collecting terms leads to:

$$w - w^{k+1} < (T - R)/(T - P)$$

And TFT remains preferred to defection when:

$$w - w^{k+1} > (T - R)/(T - P)$$

5

The Struggle for Advantage: Dilemmas of Hegemony and Competition

Nations, whether cooperating or conflicting, are presumed to act in their own interests. The situations analyzed in preceding chapters are distinguished by different constellations of payoffs, preferences, time horizons, and status quo points. But in every case states are assumed to be purposive and calculating entities that choose in line with their interests.[1]

Yet there remain alternative understandings of interest and maximization. To say that states are self-interested and self-reliant and that conflict or cooperation results from the interaction of national interests and preferences remains an incomplete explanation of either phenomenon. To say that states maximize their self-interest is also insufficient. There are various ways to define "self-interest" and "maximization," and the particular definition used by a given state has an enormous effect on both its perspectives on interactions with others and the nature of the choices it makes.

Competing Conceptions of Maximizing Power

Varying conceptualizations of the argument that states maximize their power illuminate competing notions of maximization and interest.[2]

1. A purposive-actor model can be distinguished from the more often discussed rational-actor model. An actor can be argued to be purposive and calculating even when the strict requirements of rationality are violated. The strict rational-actor model requires an actor to have a consistent, hierarchically ranked set of values, to analyze every possible option, and then to maximize in choosing that option which provides the most by way of values. Actors may not be able to analyze every option, for example, and yet still maximize within the set of options they do analyze. Such a decision may not be deemed rational, but it is based on purpose and calculation.

2. Some scholars start with the realist conception of a state's concern with maximizing

Whether it is a positive statement of what states do or a prescriptive injunction of what they must do, the basic argument that states maximize power, like the concern with power more generally, is at the heart of the study of international politics.[3] As Arnold Wolfers writes, "in its pure form the realist concept is based on the proposition that 'states seek to enhance their power.'"[4]

There are two different ways to conceptualize the maximization of power. One view, a temporal one, sees states as acting to ensure the greatest degree of power attainable over the course of time. States do what is necessary to be absolutely more powerful tomorrow than today. An alternative, cross-sectional perspective describes states as acting to be more relatively powerful than others at any single point in time.

This distinction between an absolute and relative conception of power is essential to understanding international politics, an arena in which relative power is a primary goal. Not satisfied with being stronger this year than last, states want to be stronger than their enemies at every point in time. Political leaders ask if their nations are stronger than those they consider potential challengers, not about how much stronger their states have grown over time. Nor do they ask how strong they are but if they are stronger than others whom they believe to pose a danger to them.

Important debates about military preparedness throughout history have had this relativistic character and concern. At the end of the nineteenth century, the British focused on the growth of the German fleet relative to their own and worried about ensuring a margin of naval superiority that would guarantee they could not be challenged on the high seas. On the flip side, the Germans wanted a navy to equal the British. Each nation procured weapons with an eye on the other's acquisitions. This inherently relativistic character of arms races is discussed more fully below.[5]

---

security and treat this as equivalent to a state's concern with maximizing power. This chapter discusses competing conceptualizations of the maximization of power; juxtaposing it with the previous chapter makes clear that maximizing security is not necessarily equivalent to maximizing power.

3. Unfortunately, the injunction to maximize power is often used tautologically to cover every state action from aggression to retreat. On this problem with Hans Morgenthau's argument, see Richard N. Rosecrance, "Categories, Concepts, and Reasoning in the Study of International Relations," *Behavioral Science* 6 (July 1961): 222–31.

4. Arnold Wolfers, "The Pole of Power and the Pole of Indifference," in *Discord and Collaboration: Essays on International Politics* (Baltimore: Johns Hopkins Press, 1962), p. 82.

5. Thucydides is thought of as the first international relations theorist to have discussed the balance of power because of his argument that the relative growth of Athenian power was the cause of the Peloponnesian War.

THE STRUGGLE FOR ADVANTAGE

The American competition with the Soviet Union, in all fields, has had the same relativistic character. In the 1950s Americans worried about more than the military dimension of the Soviet challenge. Rapid Soviet economic growth rates, for example, became a major source of concern as the American economy grew at a comparatively sluggish pace. The specter of the Soviet gross national product's overtaking and surpassing that of the United States haunted many American analysts.[6] Soviet scientific and technological achievements were also viewed with alarm, for these successes, interpreted as signifying that the Russians did a better job of training their young, implied also that the USSR had the potential to outstrip the United States militarily. The Soviets' ability to launch Sputnik in 1957 did not merely illuminate a danger that the Russians could develop intercontinental ballistic missiles when the United States could not yet do so; it symbolized an apparent American failure to keep up with the Russians scientifically. That Soviet lead suggested, in turn, that the United States would fall behind militarily. So the United States responded by establishing an extensive federal program to improve scientific training. The Cold War made possible federal aid to education, which had previously been defeated on states' rights grounds.[7]

The evolution of the American space program also demonstrated the nation's competitive relationship with the USSR. President Kennedy mobilized the nation in a space "race." But the goal constituted more than simply reaching or even landing a man on the moon. The key was to do it before the Russians. American space flight plans were adjusted to counter rumored Soviet schedules.

International competition is indeed like a race. The point is to get ahead of another nation (or nations) and to be the first to arrive at some goal. Just as sprinters are less interested in their actual running times than in getting to the finish line first, so nations focus on their relative rather than absolute positions.

The fact that states duplicate one another's efforts also illuminates the competitive nature of international politics. Successful innovation can allow a state to get a competitive jump on its rivals, who know that failure in the anarchic international system can mean the disappear-

6. For a discussion of American views, see E. Ray Canterbery, "The Great Soviet Growth Race," chapter 5 of *Economics on a New Frontier* (Belmont, Calif.: Wadsworth, 1968). For a balanced contemporary account, see John P. Hardt, with C. Darwin Stolzenbach and Martin J. Kohn, *The Cold War Economic Gap: The Increasing Threat to American Supremacy* (New York: Praeger, 1961).

7. James L. Sundquist, *Politics and Policy: The Eisenhower, Kennedy, and Johnson Years* (Washington, D.C.: Brookings Institution, 1968), chap. 5.

ance of their states. Hence nations duplicate one another's efforts because they are afraid not to, and innovations such as weapons systems,
styles of fighting, and even styles of military organization, diffuse from
one state to the next.[8] Not surprisingly, the American-Soviet relationship has come to include many areas of competition, over rates of
economic growth, technology, education, and so on.

This diffusion of successful innovation leads to a kind of international
socialization. Attempting to duplicate success, states take one another
as models. They follow one another into new endeavors and competitions. Rivalry between states leads to sameness precisely because
states fear for their existence and are concerned with relative gains.
Were it not for a concern about relative standing, states could specialize
and acquire various niches while becoming dependent in other areas.[9]
International specialization could occur to the degree that it has within
domestic societies. Instead, states continue to duplicate one another,
each attempting to be as self-reliant as possible when it comes to ensuring its military and economic survival and security. Minimally,
within the domain of national security and the constraints of resources,
states duplicate one another.

Struggles for power and international balances of power both involve
relative standing and position. States make calculations about their relative strength. They do not assess their power without reference to the
power of others.

Any debate between an absolute (temporal) view of power and a
relative (cross-sectional) one is a debate about decision criteria, about
what states do or what they should do, about whether states make
choices strictly on the basis of their own assessment of their returns
in various outcomes or whether they consider the returns to others. In
the foregoing chapters, states were assumed to compare and maximize
their own returns. States looked at the payoffs of others only when it
was important to ascertain what those others might do. States with
contingent choices are understood to assess others' preferences only
in order to ascertain what those others might do. Even when a state

8. This creates a methodological problem. Since innovations diffuse, their occurrences
in different states cannot be compared as if they were independent observations. Some
original cause may lead to the emergence of some phenomenon in one society, but that
phenomenon then spreads to other societies independent of the existence of the original
trigger in those other societies. This problem of controlling for diffusion has been widely
studied in anthropology, where it is called Galton's problem.

9. For such a view of international politics, see Richard Rosecrance, *The Rise of the
Trading State: Commerce and Conquest in the Modern World* (New York: Basic Books,
1986).

considers others' payoffs to determine whether they can be driven out of the game, it is understood to do so because of the returns it can reap for itself. In short, states' choices are driven solely by their own returns.

An absolute conception of self-interest is often dubbed an individualistic or egoistic one.[10] Such a formulation, which sees actors as interested only in their own returns, presumes them to be indifferent to others' payoffs.[11] They compare outcomes solely with an eye on their own payoffs and remain indifferent to the payoffs to others in those outcomes. Others' payoffs are assessed only to ascertain what they might do and only when those actions become what one's own decisions are contingent upon.

Outcome versus Utility

This distinction between decision criteria is lost in studies of strategic interaction that conflate outcome and utility. The story classically used to depict the prisoners' dilemma illustrates the straightforward translation of outcomes into preferences. In that tale the district attorney presents two prisoners each with two options and tells them what the outcomes associated with their decisions will be. The game is presented as one in which the outcomes can be clearly ranked. Yet there is no distinction made between the actual payoffs for the prisoners and the utilities or disutilities that they derive from those payoffs. The implicit assumption is that the outcomes speak for themselves. Who would not prefer to go free rather than hang? Whatever utility is associated with freedom and whatever disutility is associated with hanging, the former is clearly preferable. Hence the analyst has no need to deal with the issue of how the prisoners assess outcomes.

As in economics, analysts of world politics assume actors to be utility maximizers. Individuals determine utilities, and they assign value, differently. Economists have long known, for example, that utility need not equal money, either absolutely or in relative amounts. They recognize that an increment of one dollar means more to a pauper than to a rich person, and they also know that just as people exchange dollars for goods, they also defer income for leisure time or more mean-

10. Robert Axelrod, "The Emergence of Cooperation among Egoists," *American Political Science Review* 75 (June 1981): 306–31.
11. Such a view is at the heart of modern economics, where preferences about others' payoffs are described as "meddlesome preferences." In addition, moral philosophers dispute the legitimacy of accepting "external preferences."

ingful work. In the analysis of international relations, a state's utility is assumed to derive from its own outcomes. Hence in the analysis of strategic interaction, either outcome and utility are treated synony- mously, or one is seen as a linear transformation of the other (i.e., the rankings are comparable).

Actors may make different evaluative judgments when confronted with a single set of outcomes, depending on whether they derive their views of self-interest from a focus on relative or absolute criteria. Psy- chologists Harold Kelley and John Thibaut distinguish between a *given* matrix, which delineates the outcomes as given by the environment and the properties of the actors, and an *effective* matrix, which becomes the basis for choice. The given matrix undergoes a transformation to become the effective matrix. The effective matrix is the one that the actors perceive and the one upon which they act. As Kelley and Thibaut argue from their long study of interpersonal relations, "there is no close causal nexus between the *given* matrix and the behavior it elicits." Rather, actors respond to patterns in the given matrix to generate the effective matrix, "which is then closely linked to their behavior."[12]

Actors' orientations, whether they maximize relatively, absolutely, or jointly, transform a given matrix, a set of actual outcomes, into the effective matrix, the perceived returns from the outcomes. These ori- entations provide the decision criteria and define self-interest, and these transformations are essential to explaining the actors' choices and the patterns of cooperation and conflict that emerge from them.

One cannot infer what decisions an actor will make—nor, therefore, outcomes—merely from an assessment of the actual payoffs associated with various strategies. The existence of cooperation and conflict, pat- terns of states' behavior, and outcomes all depend not only on the payoffs but on the decision criteria that actors bring to bear in evalu- ating different situations. Actors may make decisions that appear to be irrational to anyone analyzing the given matrix but that become fully rational given the effective matrix the actors are assessing. Hence it is important to know not only actual payoffs but also the orientations actors bring to viewing those payoffs.

Competing Conceptions of Maximizing Wealth

Because they apply also to maximizing wealth, alternative conceptual- izations of maximizing power can be illustrated with arguments about

12. Harold H. Kelley and John W. Thibaut, *Interpersonal Relations: A Theory of Inter- dependence* (New York: John Wiley, 1978), p. 17.

the behavior of firms. The classic view is that firms act to maximize their profits. A corporation wants to make more and more money, wants its profits to grow and grow, and focuses on improving efficiency and growth. Alternatively, the firm may concentrate on seeking rents, on effecting a redistribution of existing sales. This can be done by lobbying to obtain preferential and beneficial government regulations. In addition, firms can choose between focusing on absolute rates of return and on market share. In the latter view, companies want to grab an ever bigger portion of the market.[13]

Each of these competing views of corporate behavior has important and startling consequences. In a world in which firms act to maximize profits, competition is not direct, and the success or failure of a firm is independent of the absolute success or failure of others. It is possible for all firms in an industry to post profits in any year, to post record profits in a particular period, and to post declines and losses in a specific period. If competition is for market share or if firms act as rent seekers, however, one firm's success comes at another's expense and one's failure must redound to another's success. Redistribution gives to one firm at the expense of another. Moreover, the market for any product may be growing, but the market total always remains at 100 percent. A firm may increase its profits in a growing market but still grab a smaller share of that market. Its success or failure may be determined by the goals it has or the criteria used to judge performance. The criteria may also determine choices. Firms may decide to forgo greater profits in order to capture a larger share of the market.[14]

These different formulations of maximizing wealth can be applied to nations as well as to firms. States maximizing absolute wealth make temporal comparisons with their past performance. They focus on the annual growth rates of their gross national products, for example. Alternatively, if their concern is with relative wealth, they worry about their growth relative to other states.[15]

13. In many industries the competition is presumed to be about market share. The battle between Coke and Pepsi provides a good example. In the 1980s these two giants engaged in a competition over acquiring other companies, the point of which was to maximize their relative resultant market share.

14. This is a conventional distinction made between American and Japanese firms in the 1980s. See James Abegglen and George Stalk, Jr., *Kaisha: The Japanese Corporation* (New York: Basic Books, 1985), p. 177.

15. In addition, states can themselves be seen as rent seekers. For an argument about the impact of rent seeking in the international political economy, see Mark R. Brawley, "Challenging Hegemony: How Cycles of Hegemony, Hegemonic Decline, Major War, and Hegemonic Transitions are Linked," Ph.D. dissertation, University of California, Los Angeles, 1989.

Consequences of Competitive Decision Criteria

These formulations are related to the difference between constant- and variable-sum games. In the former, there exists a fixed and constant reward for which players compete. Such competition is direct and conflictual, because improvement for one actor can come only at another's expense. Such instances of pure conflict contrast with variable-sum games, in which players compete for a variable good and in which improvement for one can be unrelated to another's returns. In the past, scholars questioned whether it was appropriate to view international politics as a constant-sum game of pure conflict instead of a variable-sum phenomenon in which cooperation was possible and perhaps even desirable.

Some interactions, such as territorial competition, are necessarily constant-sum because the amount of the payoff is fixed. The earth's land mass changes so slowly that one nation's control of appreciably more land must come at another's expense.[16] Territorial and boundary disputes are thus constant-sum. The acquisition of some territory by a nation will come at another's loss.

Other forms of competition are not constant-sum, per se. The amount of wealth is not fixed, so all nations can prosper simultaneously. This distinction, and that between absolute and relative wealth, is captured nicely by the late eighteenth- and early nineteenth-century debate between mercantilists and liberals. Mercantilists maintained that global wealth was fixed and that one nation's wealth increased only at the expense of others.[17] Economic liberals attacked this view and argued that because total global wealth could and did change, nations' performances were, if anything, positively correlated. Improvement for one meant improvement for others, perhaps for all. Wealth in others' hands meant demand for one's own goods and services and an increase, therefore, in one's own wealth.

Scholars' views of economic competition have changed since the eighteenth century, and the liberals' variable-sum view of economic exchange is now widely accepted. But the fact that total wealth fluctuates, that there is no fixed level of wealth, and that all can become richer and all become poorer, does not mean that economic competi-

---

16. It seems safe to ignore territorial accretions through landfills, lava flows, and naturally changing coastlines. Note that the same argument explains why land reform generates more conflict than a variety of income redistribution schemes.

17. Jacob Viner, "Power versus Plenty as Objectives of Foreign Policy in the Seventeenth and Eighteenth Centuries," *World Politics* 1 (October 1948): 9.

tion cannot be constant-sum. It means only that because wealth is not a fixed good, interactions regarding it need not necessarily be constant-sum.

The adoption of a market-share, rent-seeking, or relativistic conception of the outcome serves to transform a variable-sum game into a constant-sum one. If states adopt the market-share perspective of some corporations, international economic interactions can also become constant-sum. Although world and American manufacturing and exports have all grown dramatically in the post–World War II period, it is not difficult to find economists who point with alarm to the declining American shares of both world manufacturing and world exports in manufacturing. Because they focus on maximizing the share of manufactured exports, they transform a variable-sum issue into a constant-sum one.[18]

In short, conflict and direct competition (constant-sum games) can evolve in two ways. Interactions regarding goods that are inherently fixed, such as territory, are necessarily constant-sum. On the other hand, inherently variable-sum games can be transformed into constant-sum ones if the actors adopt relative conceptions of self-interest and competitive rather than individualistic orientations. This is the lesson of the corporate desire for profits as opposed to that of corporate competition for market shares.

The nature of international politics is thus a function of both the nature of the good in question and the orientation of the states involved. Territorial competition is necessarily constant-sum. Hence imperial competition between the major European powers at the end of the nineteenth century was fraught with conflict and quite dangerous. Latecomers to the imperial game either had to make due with colonizing territories not yet claimed or had to obtain lands from other major powers.

The constant-sum character of the major-power rivalry over colonies provides the basis for Lenin's explanation of World War I in particular and wars between capitalist states in general. Lenin argued that capi-

---

18. A concern with rank is one manifestation of relative assessments. Nations and firms concerned with being number one rather than with their actual profits care about their relative standing. Only one can be number one. An example of an emphasis on rank even in a variable-sum game is evident in some facets of the American preoccupation with Japan and is best exemplified by the title of Ezra F. Vogel's book, *Japan as Number One: Lessons for America* (Cambridge, Mass.: Harvard University Press, 1979). A focus on relative market share is a focus on rank. Economists who do mention concerns with rank refer to them as concerns with status. Martin Shubik, "Games of Status," *Behavioral Science* 16 (March 1971): 117–29.

talist states required such outposts as a direct result of their advanced economic development. But those that developed late would not be able to obtain the colonies they needed, because the world would already have been carved up. Since he believed that peaceful redistribution of colonial territories between capitalist great powers was impossible, he predicted war.[19] The end of the European colonial phase and the acceptance of decolonization defused a major area of rivalry and competition among states.

A shift from territorial to economic competition would, according to many scholars, increase interdependence and reduce conflict. Yet because nations continue to act in accordance with mercantilist formulations (in contradistinction to liberal ones), it appears that a focus on wealth is no guarantee of an absence of rivalry, competition, and hostility. National objectives to maximize the market share of international trade transform a variable-sum issue involving both common and divergent interests into a constant-sum one dominated by divergent interests.[20]

To understand the nature of international interaction, therefore, it is important to ascertain not only the objective nature of the good in question but also the decision criteria that states adopt. A competitive orientation focused on relative returns transforms the nature of competition.[21]

19. For Lenin's argument, see V. I. Lenin, *Imperialism: The Highest Stage of Capitalism* (New York: International Publishers, 1939). For criticism of his argument, see Kenneth N. Waltz, *Man, the State, and War: A Theoretical Analysis* (New York: Columbia University Press, 1959); Waltz, *Theory of International Politics* (Reading, Mass.: Addison-Wesley, 1979). For a reassessment of Lenin's view of World War I, see Steven Louis Isoardi, "Class Structure and Diplomatic Aims: On the Boundaries of International Relations in the Imperialist Epoch," Ph.D. dissertation, University of California, Los Angeles, 1986.

20. Debates in international political economy focus not only on the decision criteria that states do and should use but also on the payoffs of the various strategies they employ. Free trade, most agree, maximizes global wealth and is a dominant strategy for any state interested in maximizing the joint wealth of itself and others. In contrast, a hegemonic economic power interested in maximizing its wealth relative to others will find a protectionist strategy to be dominant. There is, on the other hand, a dispute about the appropriate economic strategy for a hegemonic power interested in maximizing its absolute wealth. Some argue that free trade is a dominant strategy for such a state. Others argue that protectionism is. Finally, still others argue that there is no dominant strategy for such a state. Rather, free trade, the first-best option, is optimal only if others pursue it as well; if others depart from free trade, the hegemon finds protectionism, its second-best strategy, to be optimal. Unfortunately, this theoretical debate is conflated with the separate issue of which decision criteria states use. See David A. Lake, "Beneath the Commerce of Nations: A Theory of International Economic Structures," *International Studies Quarterly* 28 (June 1984): 143–70; John A. C. Conybeare, "Public Goods, Prisoners' Dilemmas and the International Political Economy," *International Studies Quarterly* 28 (March 1984): 5–22. See also Raymond Riezman, "Tariff Retaliation from a Strategic Viewpoint," *Southern Economic Journal* 48 (July 1981): 583–93.

21. Competitive orientations can also be instrumental. First, they are an indicator of

With this understanding, debates about international politics can be recast. Both those who argue that aspects of international politics are inherently constant-sum and those who hold that international politics may be inappropriately viewed as constant-sum are right. Territory is constant sum. Maximizing global wealth remains a variable-sum game as long as actors take individualistic egoistic stances regarding their returns. On the other hand, variable-sum games can be transformed. Implicit in recommendations for such so-called beggar-thy-neighbor economic policies as devaluation, for example, is the assumption that since someone will be beggared, better thy neighbor than thyself.[22]

Much the same can be said about arguments regarding power and international politics. The realist view of international politics holds that states act to ensure their survival and security by maximizing power.[23] The question arises, therefore, whether power is inherently a constant-sum or a variable-sum good.

The study of international relations concentrates on those states that are great powers during a given epoch. The comparisons are invariably cross-sectional. It does not matter that many nations today are richer and more powerful militarily than, say, the Netherlands was when it was a great power. Yet no one would classify these modern nations as some of the great powers of world history. What matters is relative standing at any point in time rather than over the course of time.[24]

Weapons Acquisition, Arms Races, Arms Control, and the Prisoners' Dilemma

If defined in terms of military hardware or destructive capacity, military power is a variable-sum good. Neither the number of guns nor soldiers on this planet is fixed. The global stockpile of nuclear weapons has

---

performance. Sometimes an actor cannot tell how well it is doing except by comparing itself with others, which provides a benchmark. Second, competition may be directly instrumental because payoffs reflect rank. A company's stock price may fluctuate with changes in its market share rather than in its profitability. Salaries usually reflect rank rather than performance.

22. Economists expect nations, like people, to be indifferent to accretions of wealth by others as long as their own wealth is not adversely affected. They view preferences regarding others' wealth as envy. But in international relations, there may develop adverse security consequences of others' relative economic growth.

23. Although the security dilemma implies that maximizing power can undercut security.

24. This can be readily seen in any of the scholarly classifications of great powers. See, for example, George Modelski, *Principles of World Politics* (New York: Free Press, 1972); Jack S. Levy, *War in the Modern Great Power System, 1495–1975* (Lexington: University Press of Kentucky, 1983).

grown fairly steadily since the end of World War II. The destructive potential of armaments (and their range and speed) has increased steadily over the centuries.[25] The superpowers now have the ability to destroy all life within a given radius anywhere on the planet in less than an hour. Just as nations have become richer over the past centuries, so have they developed more destructive military arsenals.

The typical scholarly analysis of the process of competitive weapons procurements as a prisoners' dilemma presumes that it is variable-sum. The argument runs as follows: When given the choice of whether or not to arm, all states find that acquiring weapons is a dominant strategy. If other states do not arm themselves, the state that has done so can then take advantage of them. Alternatively, if others procure weapons, the state is better off arming itself for protection against exploitation. In short, procurement is preferred regardless of what any other state does. Nevertheless, all states find a world of mutual nonarmament preferable to one of mutual armament. Not only do all save the money spent on weapons, goods which consume rather than reproduce capital, but all are safer when none have weapons than when everyone does.

In this way, international relations theorists explain how self-help and anarchy lead to a world of weapons rather than to disarmament. States are caught in a prisoners' dilemma in which—because each independently concludes that it is better off arming itself—all are trapped in the Pareto-suboptimal world of mutual armament. The Pareto-optimal state of global disarmament is not an equilibrium one; it is susceptible to cheating and defection by all. To forgo weapons, nations must be assured that all others will do so as well. Moreover, they must be convinced that monitoring will keep others from betraying the agreement and catch any that do. In short, each must know that a surprise defection by another will not jeopardize its security.

This analysis of the decision to arm not only explains the historic preoccupation of states with military means of ensuring security but also forms the basis for recurrent proposals for negotiations to achieve arms control. Since nations are presumed to have a common interest in controlling their mutual weapons acquisitions, arms control agreements are seen as being Pareto-superior to a world of unrestrained acquisition. Such accords are difficult to reach, however, because they require states to move away from their dominant strategies. Each side expects the other to defect because each is recognized to have an in-

25. Bruce M. Russett, *Trends in World Politics* (New York: Macmillan, 1965), pp. 8, 10.

centive to defect unilaterally. So states must consent to measures to implement their agreement; they must find mutually acceptable mechanisms for monitoring potential defection and ensuring compliance. All must know not only that defections will be spotted but that a response can be sufficiently rapid to prevent exploitation by the cheater. If the world's two nuclear superpowers agree to destroy their strategic arsenals, each must be assured that the other follows the agreement and does not secretly keep some of its weapons. In addition, each must also be convinced that if the other does break the accord and obtain a weapon, it cannot take advantage of the situation. The incentive to defect and the fear of defection are what make arms control agreements so difficult to achieve.

This view of weapons acquisition has been extended to arguments about arms races. Arms races occur when nations decide to procure additional weapons rather than maintain the status quo. Assuming that the end results of an arms race are worse than the situation that existed before the race began, these competitive weapons buildups constitute prisoners' dilemmas.[26]

Those with this perspective correctly treat weapons accumulations as not inherently constant-sum, but others do see it this way. Although the number of weapons on the planet is not fixed, nations do not always look benignly on others' acquisitions of more weapons. The United States, for example, wants its allies to procure more weapons and sees such steps as positive.[27] On the other hand, it understands increased procurements by the Soviets to be threatening. In other words, others' stockpiles directly affect the utility a state perceives in a situation. Take, for example, the U.S. position in early 1961, when the Kennedy Administration received widely varying estimates of Soviet deployments of intercontinental ballistic missiles (ICBMs). The Air Force estimated a force of 600 to 800; the CIA put the number at 450; and the Navy at 200.[28] No matter which of these numbers was accurate, the number of ICBMs in U.S. hands remained a constant. Yet knowing the correct number of Soviet missiles was essential in determining whether the United States outpaced or lagged behind the USSR. This is because

26. For a recent such characterization, see Stephen J. Majeski, "Arms Races as Iterated Prisoner's Dilemma Games," *Mathematical Social Science* 7 (1984): 253–66. See also Russell Hardin, "Unilateral versus Mutual Disarmament," *Philosophy and Public Affairs* 12 (1983): 236–54.

27. This case is discussed in the next chapter.

28. Arthur M. Schlesinger, Jr., *A Thousand Days: John F. Kennedy in the White House* (Boston: Houghton Mifflin, 1965), p. 499.

position is not perceived merely as a function of the absolute number of weapons that one has. Rather, a nation's situation is addressed in relative terms, in an assessment of relative numbers, relative capabilities, and potential consequences.

A nation does not look benignly at a rival's accretion of military power, for each increment of military power obtained by a rival effectively reduces and degrades its own. There is, then, a *negative externality* associated with the military power of a rival. A decline in a competitor's strength increases the utility of one's own forces. Thus a view of how states rank different outcomes involves an assessment of their own capabilities relative to those of their rivals.[29]

Relativistic assessments of outcomes are at the heart of the international politics of arms races.[30] The absolute payoffs of an arms race may generate a prisoners' dilemma, but if the actors analyze the situation in competitive and relativistic terms, the prisoners' dilemma will be transformed as follows (see Figure 20).

Clearly, each actor will continue to rank the outcomes in which one obtains its best and the other its worst payoff in the same order. But their assessments of mutual armament and mutual disarmament will change. If both see one actor or the other as clearly obtaining higher returns in one or the other outcome, this perception will enter each actor's calculus in making its relativistic assessments.

In this relativistic assessment, each actor still has the same dominant

29. Since negative externalities entail costs actually experienced, they would be captured in any assessment of absolute self-interest. It can be argued, therefore, that there is no need to conflate them with relative calculations and competitive decision criteria. Yet it is the very element of competition that generates negative externalities. As I point out in the next chapter, the same weapons that generate negative externalities in the hands of an enemy can generate positive externalities in the hands of an ally. The externality derives from the relationship; it is not inherent.

It is always possible to reduce competitive assessments to individualistic ones—either by emphasizing perceived externalities from others' payoffs or by emphasizing long-term absolute payoffs. Thus a focus on market share does not mean that firms are not concerned with absolute profits, only that they are thinking about the long term. But future payoffs are uncertain and so, I argue, decisions are made on the basis of current but relativistic assessments. Market share is used as the best estimate of future profits. My point here is only that the calculus of decision is relativistic.

Regardless of whether modeled in terms of absolute or relative self-interest, the basic conclusions of the book remain: the prospects for cooperation are reduced but not destroyed by a parameter that reflects the impact of the negative externality, or an emphasis on the long run, or a relativistic calculus. In other words, conflating relativistic with egoistic calculations does not accurately reflect the bases on which decisions are actually made, but also does not alter the substantive conclusions of this chapter.

30. See the classic analysis by Samuel P. Huntington, "Arms Races: Prerequisites and Results," *Public Policy* 8 (1958): 41–86.

Figure 20. Prisoners' dilemma and competitively transformed prisoners' dilemma

	Prisoners' dilemma Actor B				Competitive transformation of PD payoffs Actor B	
	Cooperate	Defect*			Cooperate	Defect*
Cooperate	3, 3	1, 4		Cooperate	2, 3 or 3, 2	1, 4
Actor A				Actor A		
Defect*	4, 1	2, 2**		Defect*	4, 1	3, 2 or 2, 3**

*Actor's dominant strategy
**Equilibrium outcome

strategy, resulting in the same equilibrium outcome. What has changed is that the equilibrium outcome is no longer Pareto-deficient. In addition, the transformed difference matrix is a zero-sum world. One's loss is another's gain—and necessarily so. If each actor assesses every outcome by analyzing the difference between the returns, and if each sees the same set of returns, the result is a world in which every outcome is zero-sum but which is not a prisoners' dilemma. There exists no common interest in achieving a mutually more desirable outcome. There cannot be such common interests between rivals making relativistic assessments based on the same information.[31]

We return now to the question of whether arms races are prisoners' dilemmas in which both actors have an interest in negotiating a more preferred outcome. In part, the answer hinges on whether the actors are making relativistic or individualistic assessments of outcomes.

The evidence suggests that actors make relativistic judgments. The whole point of becoming involved in an arms race is to establish (or undercut) a perceived difference in position. The race is run to determine who will come out ahead. These are not athletes running by themselves on an empty track to try to establish a best personal time. The action-reaction logic of an arms race is precisely that of nations responding to each step that the other makes and without indifference to its moves. The very ways that political scientists model arms races

31. Not surprisingly, Axelrod argues that actors in prisoners' dilemmas should not be envious. Robert Axelrod, *The Evolution of Cooperation* (New York: Basic Books, 1984). For an analysis of how changing players' goals from maximizing average absolute payoffs to outscoring opponents affects Axelrod's simulations, see Roy L. Behr, "Nice Guys Finish Last—Sometimes," *Journal of Conflict Resolution* 25 (June 1981): 289–300. The importance of an absence of envy is reinforced in Theodore To, "More Realism in the Prisoner's Dilemma," *Journal of Conflict Resolution* 32 (June 1988): 402–8.

capture this relativistic logic. The famous Richardson equations, for example, model each side's arsenals (or expenditures) as some direct function of the actions of the other. States are understood to respond to the actions of others because they are seen to be adversely affected by each increment to the other's arsenal. In other words, states make inherently relativistic judgments about their respective military establishments.

Arms races cannot, therefore, be prisoners' dilemmas.[32] In fact, the most that can be inferred from the foregoing analysis is that states might not care how many weapons each nation has as long as they all have equal numbers. Two rival nuclear powers might have no preference, for example, between a world in which each has five hundred weapons and one in which each has one thousand weapons. But such rivals might not be indifferent if this presumed equivalence was invalidated by another sort of utility calculation. One party might be stronger in conventional forces, and another might rely on its nuclear arsenal to even the score, in which case a world of smaller equal numbers might be seen as effectively favoring the side with superiority in conventional forces.[33]

To sustain the view that arm races are prisoners' dilemmas, one must make the following assumptions. First, equal stockpiles must result; in other words, the race cannot be won. Because neither side can establish a superior position, the competitors enter a world of equal forces at higher levels. Second, there cannot be other factors that lead either state to prefer equality at higher numbers to equality at lower numbers. Otherwise, one state will not be indifferent between the arms race and arms control outcomes but will believe itself to be disadvantaged by the latter. That assessment, of course, could form the basis for the other state's interest in arms control.[34] Third, the cost of a race (or some other factor) must be undesirable to both parties. Neither prefers either higher or lower numbers of weapons on military grounds, but both will still choose lower numbers for other reasons.

These requisites are stringent and may not typically be met. Different

32. For a qualification of this proposition, see below.
33. This is exactly the argument that is made whenever the abolition of theater nuclear forces is proposed.
34. This problem exists even for those whose only goal is nuclear stability. If the world is more stable at mutually high levels of deployment than at mutually lower levels of deployment, one cannot be indifferent between these two worlds if one places a positive value on stability. Clearly a balance in which each side has five weapons is more unstable and more susceptible to a potential "breakout" by one side and with potentially disastrous consequences than a world in which both have 500.

states with different weapons systems and different security needs, interests, and concerns may not be indifferent between equality at different levels. Indeed, it may be difficult to establish what constitutes equality. Moreover, it may be necessary for a state to experience an arms race to believe it cannot be won by either side. Further, there may not exist other costs to tip the balance toward a mutual preference for arms control over continued racing. Minimally, one clear conclusion is that not all arms races are necessarily prisoners' dilemmas open to a mutually preferred negotiated solution.[35]

There remains one other basis for seeing arms races as prisoners' dilemmas susceptible to negotiated agreement between competitive rivals. The foregoing discussion presumes that actors are comparably concerned with their own and others' payoffs. In effect, they attach the same weight to both their own payoffs and those of others in making their assessments. When both actors take a competitive and relativistic view, and when both have the same view of the payoffs, the result is a zero-sum game.

But actors may not maximize the difference between their payoffs. They may attach different weights to their own and others' payoffs. They may be more concerned with minimizing others' gains than with maximizing their own. Alternatively, they may be more concerned with their own gains than with minimizing those of others.

Moreover, in any strategic situation, the weights attached to payoffs by two actors may differ, and this can be the basis for negotiated agreements between rivals. The first matrix illustrated in Appendix 1 is a prisoners' dilemma in absolute payoffs. The actors, concerned with absolute payoffs, would have an interest in a negotiated outcome that brought them to the Pareto-superior but nonequilibrium outcome. But the actors are presumed to make relativistic comparisons. In addition, they attach different weights to each other's payoffs. In Appendix 1, actor A maximizes the difference between its own payoffs and K times the other's payoff, whereas actor B takes the difference between its own payoff and L times the other's payoff.[36] The deductive conclusion is that

---

35. George W. Downs, David M. Rocke, and Randolph M. Siverson, "Arms Races and Cooperation," *World Politics* 38 (October 1985): 118–46. In this article the authors explicitly recognize that the outcome of an arms race is consistent with games other than the prisoners' dilemma. I find their categorization of the outcomes that they find consistent with an arms race to be too broad, however. Also see Thomas C. Schelling, "A Framework for the Evaluation of Arms Proposals," in *Frontiers in Social Thought: Essays in Honor of Kenneth E. Boulding*, ed. Martin Pfaff (New York: North-Holland Publishing, 1976), pp. 283–305.

36. This formulation differs from that developed by Joseph Grieco, "Realist Theory

when K exceeds 1/L (or L exceeds 1/K), the situation, despite both actors' emphasis on relative payoffs, remains a prisoners' dilemma in which both can be better off by negotiating and arriving at a superior outcome of mutual cooperation.

Hence negotiated arms control is a viable solution to an arms race only if one state is willing to accept an inferior position. More precisely, one state demands a relative difference that is less than the inverse fraction of that relative difference demanded by the other power. In effect, negotiations to end a bilateral arms race in which the states are concerned with relative position are possible only if one state is willing to accept relative marginal inferiority.

Arms Control Negotiations

Even when those assumptions are met and an arms race is a true prisoners' dilemma, the very process of negotiation, and the requisites of the negotiated outcome, differ from normal prisoners' dilemmas. In the typical version of the game, both sides find the equilibrium outcome to be Pareto-deficient. Negotiations then focus on how to move from that outcome to the preferred one and on issues of verification, monitoring, and the potential costs of defection. Moreover, new concerns arise as the struggle for relative position continues into the negotiations.

Nations that engage in an arms race pursue relative goals. Those relativistic assessments and concerns underlie the dynamics of any negotiations between them. Indeed, it should come as no surprise that the nation expecting to lose the race, wants a negotiated end to the competition. Like two boxers in a ring, the one getting pummeled wants the bell to end the round while the other would happily continue fighting. A call for discussion to end an arms race cannot always be seen, therefore, as evidence of the existence of a prisoners' dilemma, in which both sides prefer a negotiated solution to continued racing.[37]

Yet even evidence of mutual interest in discussions does not necessarily mean that the arms race in question is a prisoners' dilemma amenable to a negotiated solution. Each side may believe that it can

---

and the Problem of International Cooperation: Analysis with an Amended Prisoner's Dilemma Model," *Journal of Politics* 50 (1988): 600–624. He multiplies the difference parameter K by the difference between one's own and the other's payoff. The deduction made here about the relationship between K and L holds even if one adopts his formulation.

37. Indeed, in the transformed prisoners' dilemma shown in Figure 20, one actor still prefers an agreement to the arms race result (DD).

obtain a relative advantage through negotiation rather than unabashed racing. One may believe that it can slow down the other's efforts by calling for talks and so generating divisions in the other society and slowing the momentum of its procurements.[38] Or one nation may think that it can get the other to sign an agreement that enshrines its inferiority.[39] Alternatively, one may believe that calling for negotiations will free its hand to pursue a continued buildup.[40] In short, the mere fact that both sides are calling for discussions cannot be taken as evidence that there exists a potentially negotiated outcome that both would prefer to the status quo deriving from their unilateral actions and autonomous decisions.

The foregoing analysis suggests that arms control negotiations ought to be a rarity among nations engaged in weapons races and that neither proposals for such talks nor actual negotiations ought always to result in successfully concluded agreements.[41] Nations engaged in arms races are competitors struggling for relative advantage who may not have a joint interest in a negotiated outcome and who may use such negotiations as another means to obtain advantage. Arms control negotiations are thus more than discussions that deal with monitoring and verification. They involve more than the question of how to confront the consequences of cheating. They cover those issues, but only after confronting the more central one of the relative military positions held by both sides before talks began and after an agreement is concluded. Only if both sides find a negotiated outcome preferable to the status quo ante do these other concerns become relevant.

Competitors engaged in an arms race might very well be able to agree on certain issues. They can easily agree on excluding potential rivals, for example. Such agreements do not affect their positions relative to each other, just their positions vis-à-vis third parties. Alternatively, they might be able to agree to channel their race and define certain issues or locations as outside the parameters of their race.

Not surprisingly, arms control agreements have not ended arms

38. Commentators on the right have employed precisely this argument to explain the Soviets' use of calls for negotiations.

39. Commentators have argued that a democracy involved in arms control negotiations will make greater concessions and sign inferior bargains as a result of domestic pressures.

40. Critics on the left have so characterized the Reagan administration and its arms control negotiating stance.

41. As long as nations are in a true competitive race, the limiting conditions that would generate a prisoners' dilemma and lead to negotiated solutions would be rarely met. One nation giving up the race or the racing nations tacitly arriving at mutually acceptable ratios of relative power would seem more likely.

races, even when they have addressed the arsenals of the rival powers. The greatly heralded hot-line agreement, which established a direct link between the superpowers, for example, did not address weapons issues at all. The test-ban treaty was only a partial prohibition on testing; it ruled out aboveground tests, but not others. Despite all the difficulties in negotiating this accord, it represented no more than a mutual decision by two nations, both of which had perfected underground testing, to get rid of a form of testing that they no longer needed, that was clearly harmful, and that brought them bad publicity, in favor of a modus vivendi that included among its consequences pressure against testing by potential members of the nuclear club. Similarly, the nonproliferation agreement is little more than a decision by two members of a duopoly to restrict the entry of future members and not to assist others' attempts to join. Their own forces and arsenals remain unaffected. The mutual interest here, although clear, had nothing to do with limiting superpower arsenals. Indeed, the utility associated with these stockpiles is ensured by an agreement to prevent the entry of potential rivals.

Arms control agreements that actually did affect the arsenals of the signers (as opposed to those of would-be rivals) also constitute a record of partial and temporary accords. States engaged in an arms race struggling for relative advantage obviously want to channel the race toward their areas of strength. Determining the size and shape of the playing field is an old tactic of competitors, and sometimes their interests in this regard are mutual.[42] Both sides in an arms race may want to direct their rivalry away from a particular area. The United States and the Soviet Union signed a permanent agreement on the deployment of antiballistic missile systems, for example. The agreement did not do away with their underlying competition or with their arms race but eliminated one particular weapons system from that escalation. This instance is representative of most American-Soviet agreements, which when they actually limit arms, have invariably been partial accords requiring the exclusion of particular weapons systems. The superpowers were able to reach a consensus about intercontinental ballistic missiles because they excluded intermediate range forces, land-based forces, and bombers from the calculus. The historical record is replete with partial accords followed by continued arms racing.

42. Baseball stories about groundskeepers being instructed to prepare fields in such a way as to maximize the advantage of the home team are legion, as are those about teams moving the fences in their stadiums in order to maximize the potential of their current hitters and pitchers.

In such partial agreements, states exclude certain issues from consideration. No nation compromises in those areas in which it holds some advantage, which it perceives as being its strong suits. The opposing side, of course, wants compromises in precisely those areas. Each uses the negotiations to free its own hands while constraining its rival's. The most likely result is a partial agreement, one covering only those shared concerns for which there exists a mutually acceptable solution.

In short, arms control agreements very often do little to reduce arsenals or to reduce competition and rivalry. Such accords may well be about third parties or tangential issues. When they do address weapons systems, they are likely to be partial agreements rather than comprehensive ones, understandings that channel arms races rather than do away with them. The results are typically freezes of or rollbacks on only certain classes of weapons systems.

The series of naval conferences and accords of the interwar period illustrate the dynamics of negotiations among rivals engaged in an arms race. The United States, already the equal of Great Britain, until then the foremost naval power in the world, called the Washington Conference of 1921–22. Now easily able to outdistance Britain in any naval buildup, the United States preferred to institutionalize the extant distribution of naval power. It called the Conference, therefore, in hopes of heading off expected postwar naval construction programs among the major powers. The resulting naval armaments treaty established a ratio of 5:5:3:1.67:1.67 for the tonnage of aircraft carriers and capital ships among the United States, Great Britain, Japan, France, and Italy.[43] The accord reflected the current relative naval strength of these countries, but excluded cruisers, destroyers, and submarines, on which the negotiators could not reach an agreement. Not only the major naval powers, but the minor powers, too, were unable to agree on ratios in these classes of ships. France, for example, worried mainly about Italy and the Mediterranean. It accepted parity with Italy in principle only, because it had more aircraft carriers and capital ships and did not believe that the Italians could afford to match its fleet. On the other hand, France wanted a free hand to build lighter ships in order to solidify its superiority over Italy.[44]

43. Important political accords were reached in Washington as well. The powers recognized that an agreement on ratios of naval forces would not, in and of itself, be enough to head off their competition. Hence they also concluded agreements regarding China and their rights in insular possessions in the Pacific. The subsequent failure of the naval agreements must be seen, therefore, as occurring despite attempts to deal also with underlying political problems as well as their manifestations in naval construction plans.

44. Joel Blatt, "The Parity That Meant Superiority: French Naval Policy toward Italy at

Not surprisingly, an arms race soon began in the classes of ships not covered by the accords of 1922. As a result, the United States called for a second conference to deal with the manifest competition in cruisers, destroyers, and submarines. France and Italy declined the invitation. Although the three remaining powers met in Geneva in 1927, they could reach no agreement. Another conference took place in London in 1930. This time, Great Britain, the United States, and Japan reaffirmed the 5:5:3 ratio in big ships and agreed on a new 10:10:7 ratio for small cruisers and destroyers. The latter ratio reflected Japan's desire to increase its relative standing but deflected its hope for true parity. The Japanese, with major building programs during the 1920s, had already achieved parity with the United States, so the new ratios actually required the Americans to build new ships. The London naval agreement thus differed from the 1922 Washington accord in not reflecting the true capabilities of the three navies.[45] The Japanese agreed, in principle, to proportions that were lower than their actual relative position. When the United States signaled its intention to enhance its navy and bring it up to the limits allowed under the 1930 treaty, Japan announced its withdrawal from the accords.

The interwar experience with arms control demonstrates the inherent problems described above. On the one hand, one might have thought that the prospects for arms control appeared good. A war-weary world had reason to control the budgetary requirements of arms races. Further, the arms control agreements were conjoined with political agreements intended to resolve the underlying disputes. Yet success in achieving arms control was temporary and illusory.

The dynamics of the negotiations reflect the relative and competitive elements described above. All parties entered each set of negotiations interested in constraining their rivals' building programs. They fashioned temporary agreements (the initial 1922 accord was to expire in 1936) limited to navies—indeed to certain classes of ships. (Interwar attempts to control airplanes and land armies failed.)[46] The partial agreements then channeled the race into the categories of ships they

the Washington Conference, 1921–1922, and Interwar French Foreign Policy," *French Historical Studies* 12 (Fall 1981): 223–48.

45. Stephen E. Pelz, *Race to Pearl Harbor: The Failure of the Second London Naval Conference and the Onset of World War II* (Cambridge, Mass.: Harvard University Press, 1974), p. 2.

46. In fact one reason for the French and Italian refusal to take part in subsequent rounds of discussions was an unwillingness to deal with naval issues separate from the military balance on land. This point is made by René Albrecht-Carrié, *A Diplomatic History of Europe since the Congress of Vienna* (New York: Harper and Brothers, 1958), p. 445.

did not limit. Although a subset of the original signers managed to hammer out a second agreement to deal with a larger array of ships, it involved a disjuncture between the agreed-upon limits and the actual numbers. The Japanese retained the fleet they had built during the 1920s but accepted future American construction that would put them in an inferior position. The United States accepted extant parity in exchange for the future right to augment its forces. But when the United States exercised that right, the Japanese withdrew from the agreement. In short, the competition continued, and the agreements intended by the respective parties to contain the others' relative positions served only to channel the race into new waters. As long as the United States insisted on superiority and the Japanese on parity, there could not really be any meaningful arms control agreement.

The bleak historical record of arms control agreements affirms the theoretical expectations. Such agreements have not aged well, and as the preceding analysis would predict, they have come under domestic attack in all nations for giving rivals too free a hand while tying one's own too tightly. In an anarchic world, one in which states rely on themselves for their own security, arms control agreements, like arms races, are judged by their effect on the relative military standing of the competitors.

Competition and Negotiated Agreements

This view of arms control negotiations can be extended to any talks in which the participants are competitors with relativistic concerns and interests. It has been argued that military issues differ from economic ones in that economic concerns are not necessarily constant-sum. Indeed, material self-interest is more individualistic than competitive. One's accumulation of wealth and what it brings is not degraded (unless one is a status seeker) by the wealth of others. The enjoyment of a million dollars is presumably unaffected by a rival's position as pauper or millionaire. Moreover, it is hard to see why rivalries need exist in this area.[47] After all, everyone can become richer. Short of a vindic-

---

47. Hence economists disparage envy and describe competitive calculations as constituting meddlesome preferences. It is known that conflicts between decision criteria can arise because individuals have preferences regarding others' payoffs; see Appendix 3. Further, increasing numbers of economists recognize that there can be a competitive element in economic decision making: that individuals may, for example, have preferences not just about their own income but about income distribution. For economists who address this issue, see Reuven Brenner, *History—The Human Gamble* (Chicago: University of Chicago Press, 1983); Brenner, *Betting on Ideas: Wars, Inventions, Inflation*

tiveness that leads someone to perceive success as lessened by others' gains and increased by their poverty, wealth is not a constant-sum game and need not be seen as one. Hence negotiations to liberalize international exchanges, so that all can benefit, ought to be straight-forward. The ease of reaching an accord should be increased by the fact that defection can be readily verified and the consequences of cheating are not particularly great. A state that violates a trade agree-ment and increases trade barriers, for example, cannot jump ahead and become a threat in the way that one that violates an arms control agreement can. In the trade arena there is plenty of time to respond to violations, and the consequences of defection are hardly of the same magnitude as those of defection from a military agreement, which may endanger a state's survival.

Even trade agreements, which presumably are more readily achieved, have often been difficult to negotiate because of competitive concerns. Just as firms compete over market shares and compare degrees of profit-ability, so do nations. Nations focus not only on the absolute gains from trade but on relative ones as well.[48] They may be wary of agreements that hold out greater returns for others than themselves. They may find no comfort in their own absolute increases if others obtain still more. Even here, then, negotiations can come to involve the division of the gains. It is this concern with the relative implications of economic agreements that has been a stumbling block even in the area of maximizing wealth.

In fact, the struggle for relative gains provides the background for the two types of trade disputes. One kind of trade conflict is between po-litical rivals who do not trade much with each other because neither wants to become dependent on the other (they have political reasons to avoid accepting any inter-nation division of labor). Nor does either want the other to become enriched by the gains that accrue from trade and exchange. What evolves is the economic warfare of embargoes, sanctions, and export controls. Here trade conflict ensures little by way

---

(Chicago: University of Chicago Press, 1985); Brenner, *Rivalry: In Business, Science, and among Nations* (New York: Cambridge University Press, 1987); and Robert H. Frank, *Choos-ing the Right Pond: Human Behavior and the Quest for Status* (New York: Oxford University Press, 1985). See also Stefan Valavanis, "The Resolution of Conflict When Utilities Interact," *Journal of Conflict Resolution* 2 (June 1958): 156–69; and Harold M. Hochman and Shmuel Nitzan, "Concepts of Extended Preference," *Journal of Economic Behavior and Organi-zation* 6 (1985): 161–76. Also note the brilliant contribution by Fred Hirsch, *Social Limits to Growth* (Cambridge, Mass.: Harvard University Press, 1976), and his conceptualization of positional goods.

48. "A state worries about a division of possible gains that may favor others more than itself" (Waltz, *Theory of International Politics*, p. 106).

of exchange, at least nothing that can be seen as strategically important. Such trade conflict reflects a willingness to forgo gains from trade so that others might not also profit.

Classical trade wars, on the other hand, break out when one state attempts to change the relative balance of returns from the exchanges between nations. In such cases, the states involved tend to be important trade partners that have a history of extensive commercial exchange, that have allowed some division of labor to develop between them, and that have agreed to previous trade accords. Still, conflicts arise because of struggles for relative advantage within the context of ongoing exchanges. Trade wars follow an attempt to change the nature of both the extant agreement and the evolution of exchange; this is exemplified by the trade wars of the late nineteenth century, which all erupted during renegotiations of trade agreements.[49] In this way, trade wars resemble those labor strikes that result from the inability of workers and management to negotiate a new contract. The negotiations over new terms by which to continue an old relationship involve a struggle over advantage as each side tries to obtain the best bargain possible. A strike or lockout represents the means by which one party or the other uses its presumed bargaining strength to effectuate a better deal. Trade wars are similar.[50]

When Competition and Individualism Conflict

An actor's orientation, its emphasis on relative or absolute position, affects how it assesses a given situation. Competitors and individualists may very well act differently in similarly structured circumstances. A competitive or an individualistic orientation may be the critical determinant of whether conflict or cooperation results.

Whether and how orientation matters varies with the context, how-

49. See my discussion in Arthur A. Stein, "The Hegemon's Dilemma: Great Britain, the United States, and the International Economic Order," *International Organization* 38 (Spring 1984): 355–86.

50. Joseph Grieco has stressed the importance of the actors' emphasizing relative gains even when economic exchange is mutually beneficial. But he does not distinguish between relative gain that ensures no area of agreement (i.e., the case of pure competition discussed above) and the best possible deal in the context of a mutually desired exchange. In the latter case, relativistic concerns can generate conflict but should not prevent agreement. There is a difference between competition that generates pure conflict and disagreement about how to divide the gains (what economists call the surplus) from trade and exchange. See Joseph Grieco, "Anarchy and the Limits of Cooperation: A Realist Critique of the Newest Liberal Institutionalism," *International Organization* 42 (Summer 1988): 485–507.

Figure 21. Different dominant strategies for different decision criteria

	Actual payoffs Actor B			Relative gains Actor B	
	Option 1	Option 2*		Option 1	Option 2*
Option 1	15, 2	12, 3	Option 1*	13, −13	9, −9
Actor A			Actor A		
Option 2*	25, 23	30, 29	Option 2	2, −2	1, −1

*Actor's dominant strategy

ever. In the prisoners' dilemma, the nature of an actor's orientation does not change its dominant strategy. As Figure 20 makes obvious, defection is a dominant strategy in both the prisoners' dilemma and the competitively transformed PD. What changes with the altered situation is that the dilemma disappears. In the PD, the dilemma arises because the equilibrium outcome is Pareto-deficient; both actors prefer the outcome of mutual nondefection. When the two actors appraise the PD with competitive orientations, however, they eliminate this element of mutual interest. Indeed, competitors who assess every outcome in relativistic terms find few common interests. A competitive orientation transforms the PD into a situation in which mutual defection remains an equilibrium outcome, but one that is no longer Pareto-deficient: one actor finds cheating preferable to mutual nondefection.

Of particular interest to the study of international politics are those situations in which different orientations generate different choices and strategies. Above is a hypothetical situation in which the cell entries represent both actual and relative units of return to the two actors (Figure 21).

Actor A clearly maximizes its individualistic self-interest by choosing option 2. Option 2 is a dominant strategy for actor A in that it maximizes the absolute number of units it will obtain. If, on the other hand, actor A is concerned about relative position, option 1 is its dominant strategy because it provides a greater margin of superiority than option 2.

Actor A is caught in a competitor's dilemma. Recognizing in both situations that option 2 is actor B's dominant strategy, actor A knows it should choose 2 to obtain 30 points and maximize its own wealth. Yet doing so will provide B with 29 points. On the other hand, given that B will choose option 2, A will get only 12 points by choosing option 1. Yet that outcome provides B with just 3 points, giving A a margin of 9 as opposed to a margin of only 1.

Actor A confronts the dilemma of having to choose between its relative margin over B and its own absolute wealth. Actor A multiplies its returns by two and a half in choosing option 2 over option 1. At the same time, however, it multiplies B's returns in the process almost tenfold.

Situations in which different strategies emerge as dominant for individualistic and competitive orientations are those in which the actor with the dilemma has a greater impact over the other's returns than its own.[51] That is, it has greater fate control than reflexive control.[52] It is better positioned to help or hurt another than itself. It must choose between absolute wealth and relative standing.[53]

I have argued elsewhere that this is the problem confronting hegemonic powers. Despite their great ability to structure the rules of the game, they often confront the dilemma that their actions have a greater impact on others than on themselves. They must often choose between absolute and relative standing. The liberal economic decisions that economists describe as adding to the wealth of all typically endow the impoverished more than the already rich.[54] A hegemonic power's decision to enrich itself is also a decision to enrich others more than itself. Over time, such policies will come at the expense of the hegemon's relative standing and will bring forth challengers. Yet choosing to sustain its relative standing, given the structure posited here, is a choice to keep others impoverished at the cost of increasing its own absolute wealth. Maintaining its relative position has obvious costs not only to others but to itself. Alternatively, maximizing its absolute wealth has obvious benefits but brings even greater ones to others.[55]

The United States confronted this kind of situation following World War II, when it stood atop the world's hierarchy of nations with un-

51. The formal proof is in Appendix 2.
52. Kelley and Thibaut, *Interpersonal Relations*.
53. This presumes that the issue is not inherently constant-sum. With territorial disputes, as with any other inherently constant-sum issue, any strategy that is dominant for absolute payoffs is also dominant for relativistic assessments.
54. This is a hotly debated proposition among development economists and dependency theorists. I am not prepared to argue that liberal policies always undercut relative standing. I am ready, though, to argue that there are circumstances in which they do, and in such cases, the rich and powerful confront a difficult choice.
55. Assuming that it is their choice and that they can make it stick, the wealthy confront much the same dilemma in opting to institute certain redistributive domestic economic policies. Such policies saved capitalism during the New Deal and increased the possibilities for accumulating wealth by providing the capital that sustains demand. Yet they also reduce societal inequalities and create new wealth holders. Old money can sustain its relative standing, but only at the cost of others' continued immiseration and at the cost of foregone realizable absolute wealth.

paralleled national wealth and power. Much of the rest of the industrial world stood in ruins, destroyed or exhausted by war. A policy to maintain this relative position may or may not have succeeded in the long run, but the United States could certainly have allowed the European and Japanese recoveries to be protracted. Instead, the United States chose to hasten that recovery and enrich itself at the same time. American policymakers argued on behalf of the Marshall Plan and other American programs that maintaining the nation's exports required putting dollars in others' hands.[56] European and Japanese recovery would provide American producers with an immensely greater market. No one else could supply what U.S. producers could. Foresight should have made it clear, and to some it did, that such policies would generate relatively greater growth elsewhere than in the United States and create future economic competitors, if not necessarily political challengers.[57]

Competitive Payoff Assessments and Chicken

Relative assessments are made not only because actors are interested in maximizing relative returns, but because comparing their payoffs helps them decide what to do. This is best illustrated using the game of chicken.

In the game of chicken two actors each have contingent strategies and a mutually least-preferred outcome. The game is competitive and has two equilibria, each of which is one actor's best but the other's second-worst. There is also a dimension of mutual interest, for they agree on a second-best outcome and on a least-preferred one. The result is competitive risk taking, with each actor attempting to preempt the other by being the first to establish that it will defect, for in this way it forces the other to cooperate. In its ordinal form, the game is symmetrical; each actor's preference ordering mirrors the other's. As a result, they have the same incentives, and the forces under which they operate are identical. Each wants to defect and have the other cooperate. Each must cooperate if it believes that the other will defect. Each has an incentive to convince the other of its determination to defect under all circumstances. In other words, each has reason to deceive (discussed in Chapter 3).

56. Other arguments on behalf of such programs related to communism. Yet these arguments, too, implicitly maintained that the United States was in a position to enrich others very rapidly and had an interest in doing so that was separate from its interest in enriching itself.

57. See my "The Hegemon's Dilemma."

Figure 22. A virtually zero-sum game of chicken

	Actual payoffs Actor B			Rank ordering Actor B	
	Cooperate	Defect		Cooperate	Defect
Cooperate	10, −10	−25, 25	Cooperate	3, 3	2, 4
Actor A			Actor A		
Defect	25, −25	−500, −1,000	Defect	4, 2	1, 1

Yet the ordinal form of the game encompasses a variety of constellations of real payoffs. It is possible, for example, for a game of chicken to be composed only of positive payoffs or, on the other hand, for it to be a situation in which all but one of the payoff combinations is zero-sum. In Figure 22, three of the four cell entries are zero-sum, but the game is one of chicken because of the fourth cell, the one that contains the mutually least-desirable outcome. Indeed this situation is a particularly conflictual and competitive game of chicken. Further, the payoffs associated with the worst outcome for the two actors are different. *Reversed on p. 72-3*

The existence of nuclear weapons transforms crises between nuclear superpowers into games of chicken. First, the existence of nuclear arsenals and intercontinental delivery systems renders a nuclear war the worst outcome for both parties, even though their losses in a nuclear war may not be the same. Moreover, once a crisis starts, however it may have been triggered, any resolution involves one's loss and the other's gain. The Cuban missile crisis provides an example. If the United States had capitulated to the Soviet emplacement of missiles in Cuba, the USSR would have been the big winner and the United States the big loser. In actuality, the Soviets' decision to retreat in the face of American pressure represented a defeat for them and a victory for the United States. Any form of mutual compromise would either have left one party better off at the other's expense or would have left the position of both unchanged (i.e., both would get zero). The unique element of nuclear crises is the especially disastrous consequences for both actors of a nuclear war, the outcome of mutual defection. That the consequences of the least desirable outcome may be asymmetric and more costly to one actor does not matter in the ordinal version of the game.

Yet actors in a game of chicken do not ignore the actual payoffs. Indeed there is a substantial literature suggesting how numerical payoffs should be assessed in order to decide what to do in such a situation of contingent choice. Daniel Ellsberg, lecturing on "the theory and practice of blackmail," argued that the decision in such a case depends on one's "critical risk," a threshold representing the maximum risk of punishment an actor is willing to accept.[58] Ellsberg argued that any state can concede or remain firm when it acts in sequence. Concession provides the payoff associated with capitulation. Standing firm entails uncertainty, because the outcome depends on whether the other actor then concedes or stands firm in its turn. Hence a state can compare the payoff of capitulation against an expected utility to be derived from standing firm. The highest probability a state can attach to another's standing firm while itself remaining indifferent between conceding and standing firm is its critical risk. In deciding whether to cooperate or defect, an actor compares its critical risk against its assessed likelihood that the other will remain firm. Each state derives its own critical risk and compares this against its perception of the other's firmness. But if each actor has precise knowledge of its own and the other's payoffs, then each can calculate the other's critical risk, and the state with the lower critical risk will capitulate (Figure 23).[59]

Such a comparative assessment of critical risk is equivalent to assessing who has more to lose. If the payoffs for the two actors are

58. Daniel Ellsberg, "The Theory and Practice of Blackmail," in *Bargaining: Formal Theories of Negotiation*, ed. Oran R. Young (Urbana: University of Illinois Press, 1975), pp. 343–63. For applications of critical risk to the game of chicken, see Glenn H. Snyder and Paul Diesing, *Conflict among Nations: Bargaining, Decision Making, and System Structure in International Crises* (Princeton, N.J.: Princeton University Press, 1977); Robert Jervis, "Bargaining and Bargaining Tactics," in *Coercion, Nomos*, vol. 14, ed. J. Roland Pennock and John W. Chapman (Chicago: Aldine Atherton, 1972), pp. 272–88. For a discussion contrasting two measures of critical risk, see R. Harrison Wagner, "Deterrence and Bargaining," *Journal of Conflict Resolution* 26 (June 1982): 329–58. Also see Barry O'Neill, "A Measure for Crisis Instability with an Application to Space-Based Antimissile Systems," *Journal of Conflict Resolution* 31 (December 1987): 631–72.

59. Snyder and Diesing use Ellsberg's notion of critical risk and avoid the issue of interpersonal comparisons by assuming that each actor knows its own payoffs but not those of its opponent. Each actor compares its critical risk, derived from its own payoffs, against some general assessment of the other's firmness. If, on the other hand, actors knew one another's payoffs, they would compare one another's critical-risk levels. Further, one way to generate an assessment of the other's resolve is to generate payoffs and calculate the other actor's critical risk. In short, although we may assume that actors do not know one another's actual payoffs, using critical risk as a means of deciding what to do entails a comparative assessment of payoffs. Indeed Snyder and Diesing do use relative critical risks to assess relative bargaining power; see *Conflict among Nations*, p. 191. Further, although referring to each state's perception of the other's credibility in threatening to stand firm, they note that credibility is "logically equivalent to estimating the other's critical risk" (p. 192). Robert Jervis argues that states can take steps to manipulate the payoffs of others in order to change relative critical risks; see Jervis, "Bargaining and Bargaining Tactics."

Figure 23. Critical risk

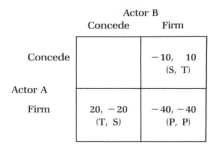

Critical Risk = T − S/T − P, where T = temptation > S = sucker > P = punishment
A's critical risk = .5
B's critical risk = .6

symmetrical (except that the worst outcome for each is different), the
state that has more to lose from that outcome has the lower critical
risk and will capitulate. Such comparisons are reflected in assessments
of who's "got more to lose."[60] Whether one party will lose more than
another should not matter. If the situation is analyzed in ordinal terms,
it makes no difference whether the mutually least desirable payoff is
−500 to both players or −500 for one and −2,000 for the other. In
either case, we expect that one will back down. To the players, however,
it does matter, for the question of who has more to lose in a confron-
tation becomes the basis for ascertaining which will escalate and signal
its determination to defect given its expectation that the other will
capitulate. Such talk is prevalent in trade wars and nuclear escalation.

One view of strategic deterrence, for example, holds that a state's
ability to deter rests on its ability to escalate its threats in a nuclear
crisis. And this, in turn, requires that it be able to inflict more damage
in any nuclear exchange, regardless of where on the ladder of nuclear
escalation such an exchange might occur.[61] Hence, as nuclear stockpiles have grown and become increasingly sophisticated, the essence
of deterrence has become equated, for some, with the ability to escalate
and "win" at every level of escalation. This is referred to as escalation

60. This principle is captured in any number of social principles. E.A. Ross's "Law of Personal Exploitation" states: "In any sentimental relation the one who cares less can exploit
the one who cares more." W.W. Waller and R. Hill's "Principle of Least Interest" argues: "That
person is able to dictate the conditions of association whose interest in the continuation of
the affair is least." Both of these are cited in John W. Thibaut and Harold H. Kelley, *The Social
Psychology of Groups* (New York: John Wiley, 1959), p. 103.
61. The notion of market power represents the equivalent formulation in economic
issues.

dominance. In this way of thinking, therefore, deterrence requires the credible ability to escalate in tandem with an adversary, thus providing no point at which the adversary can force capitulation.[62]

The entire logic of escalation dominance is relativistic. On every step of the ladder, one side must be able to inflict damage at least comparable to that of its adversary. If it cannot, if at any level it has more to lose than its adversary, it would have to capitulate. The desire to avoid being in the position of having more to lose and the policy of building up one's forces so that it is possible at every level to ensure that the adversary loses as much (however that is measured) are at the heart of escalation dominance and the entire war-fighting school of deterrence.

Ironically, the entire dispute about appropriate deterrence doctrines hinges on the presumptive decision criteria from which the debaters begin. Those who argue that absolute levels of capabilities and payoffs are all that matter conclude that deterrence in a game of nuclear chicken is maintained by the existence of bad outcomes associated with nuclear war. Those who focus on relative capabilities and payoffs, on the other hand, conclude that it is important to be able to inflict more damage than one's rival can. The entire issue rests on whether nations and their leaders make absolute or relative assessments. In such situations, even though states are interested in maintaining deterrence and ensuring their absolute payoffs and are not seeking to maximize relative gain, they may still decide between cooperation and defection on the basis of a comparative assessment of who has more to lose.[63]

Such comparative judgments about who will have more to lose are at the root of decisions made before the emergence of crisis. American strategic decisions before the fact to ensure escalation dominance are predicated upon its presumed importance should a crisis arise. The same kind of argument can be made about national economic decisions. Diversification may not make as much economic sense as specialization intended to capture the benefits of trade, but

62. Escalation dominance is usually defined as the ability to defeat "aggression at all levels of violence, short of all-out war" (Robert Jervis, *The Illogic of American Nuclear Strategy* [Ithaca, N.Y.: Cornell University Press, 1984], p. 59). It is being used here to mean an ability to inflict higher costs than one suffers, at all levels of violence. In other words, it refers here to making certain that the other has more to lose rather than ensuring an ability to win a military victory.

63. Contrast this argument with that of Jervis, *Illogic of American Nuclear Strategy*, pp. 59–63, 126–46. Jervis argues that in a nuclear world, absolute military capabilities matter more than relative ones and that competitions in risk taking can be independent of escalation dominance. My point here is that assessments of who has more to lose can be at the heart of competitions in risk taking.

a country may choose to diversify in order to avoid exploitation in conflictual games in which the party with more to lose will have to capitulate.

Conclusion

The conflicts evident in international politics represent more than the conflicts of interest that arise in a world of scarcity and anarchy. They are exacerbated by competitive struggles for relative advantage that transform areas in which there are common interests into arenas of competition. When competition already exists, a concentration on relative returns exacerbates it. The orientations of actors, the decision criteria they use, do make a difference. Preferences reflect not only actual payoffs but also the bases of assessment. A competitive orientation creates conflict of interest where none need exist. Even if actors do not pursue relative gains, comparative assessments about such questions as who has more to lose can also affect strategic choice. But as the next chapter indicates, not all dyadic interactions are characterized by competitive orientations.

The realist emphasis on competition is partly correct. The existence of negative externalities and a concern with relative assessments can, but need not necessarily, preclude otherwise attainable cooperation. Their negative impact on the prospects for cooperation is greatest in bilateral relationships (or in dyadic rivalries). Relativistic concerns do create conflicts, but only in specific circumstances.

Appendix 1
Relative Weighted Calculations in a Prisoners' Dilemma

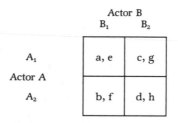

Assume the above to be a Prisoners' Dilemma:

$$\text{assume } b > a \text{ and } d > c; \quad g > e \text{ and } h > f$$

Assume competitive calculations but with parameter K for Actor A, and L for Actor B:

	Actor B	
	B_1	B_2
A_1	$a - Ke, \; e - La$	$c - Kg, \; g - Lc$
A_2	$b - Kf, \; f - Lb$	$d - Kh, \; h - Ld$

Clearly D remains dominant for both, and DD is the equilibrium outcome. Under what conditions is CC > DD? That is, can the following inequalities hold true?

$$(1) \; a - Ke > d - Kh; \quad e - La > h - Ld$$

From the original game, we know that

$$a > d \text{ and } e > h$$

If equation (1) is true, then

$$a - d > Ke - Kh; \quad e - h > La - Ld$$
$$a - d > K(e - h); \quad e - h > L(a - d)$$
$$(a - d)/(e - h) > K; \quad (e - h)/(a - d) > L$$
$$(e - h)/(a - d) > 1/K$$

That is, L and K must move inversely for the transformed competitive calculations still to generate a PD in which CC is preferred to DD by both.

Appendix 2
Implication of a Different Dominant Strategy for Different Decision Criteria

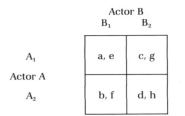

Actor B
B₁   B₂

	B₁	B₂
A₁	a, e	c, g
A₂	b, f	d, h

Actor A

If A_2 is a dominant strategy for absolute returns, then

$$b > a \quad \text{and} \quad d > c$$

If A_1 is a dominant strategy for relative gains, then

$$a - e > b - f \quad \text{and} \quad c - g > d - h$$

By algebraic manipulation:

$$(f + h) - (e + g) > (b + d) - (a + c)$$

By definition:
A's reflexive control over its own returns: $(b + d)/2 - (a + c)/2$
A's fate control over B: $(f + h)/2 - (e + g)/2$

NOTE: Reflexive control and fate control are from Harold H. Kelley and John W. Thibaut, *Interpersonal Relations: A Theory of Interdependence* (New York: John Wiley, 1978), pp. 31–43.

Appendix 3
Sen's Paradox

The domestic public policy dilemma commonly known as the "impossibility of a Paretian liberal" both involves a conflict between decision criteria and provides an example of an actor's having preferences about another's payoffs.[1] A simple story illustrates the conflict between liberalism and the Pareto principle. Two people hold different views of a particular book. The prude's (person 1) first choice is that no one read the book (option z), but the prude would rather read it himself or herself (option x, that person 1 read it) than have person 2 read it (option y). That is, the prude most prefers that all be denied, and next prefers to censor rather than be censored (on the presumption of being less susceptible to wayward influences). In contrast, person 2 abhors censorship (option z is the worst choice) and next prefers that the prude read the work (prefers x to y). In short, person 1 prefers z to x to y, whereas person 2 prefers x to y to z.

In choosing a societal assessment of the options, different values come into conflict. Some with liberal values might argue that in a choice between one person reading the book and no one reading the book, that particular person's preferences should count. Thus, in a choice between person 1 reading the book and no one reading it, person 1's views should hold (i.e., the society should prefer z to x). Similarly, person 2 should be decisive in a society's choice between person 2 reading

1. Also known as "Sen's paradox" after its formulator, Amartya Sen, the original formulation of the dilemma is in Sen, "The Impossibility of a Paretian Liberal," *Journal of Political Economy* 78 (January/February 1970): 152–57. For a review of the substantial literature that developed on this topic, see Amartya Sen, "Liberty, Unanimity and Rights," *Economica* 43 (August 1976): 217–45. Both essays are reprinted in Amartya Sen, *Choice, Welfare and Measurement* (Oxford: Basil Blackwell, 1982). The notes on pages 26–27 of the "Introduction" to Sen's volume list many articles devoted to this issue, including ones that provide other examples of the conflict. For a recent book-length treatment, see John L. Wriglesworth, *Libertarian Conflicts in Social Choice* (Cambridge: Cambridge University Press, 1985).

it and no one reading it (i.e., society should prefer y to z). Liberal values should lead to the conclusion that it is better that no one read the work than to force person 1 to read it and that it is better that person 2 read it rather than not be allowed to. In a liberal society, y should be preferred to z, and z to x. Yet the Pareto criterion that what all individuals prefer should be socially preferred is violated because both people prefer x to y. Combining liberalism and the Pareto principle generates a cycle of preferences in which each solution is inferior to some other option. Sen's paradox is that of a contradiction between two criteria of choice: the Pareto principle (which asserts the priority of unanimous preference orderings) conflicts with liberalism (defined as a person being decisive over personal matters).

The paradox incorporates a particular definition of "liberalism," one consistent with the individualistic focus on absolute payoffs discussed in this book. This definition (which allows individuals to be decisive in matters that regard them solely) effectively excludes the preferences of one person regarding the behavior of others. The liberalism criterion in the above example excluded the views of person 2 on the question of person 1's reading the book and excluded the views of person 1 on person 2's reading the book. This liberalism entails a view of self-regarding interests as the only admissable ones in the formation of a social choice. It parallels the definition discussed earlier in the chapter of a liberal view of self-interest, one that has actors looking solely at their own payoffs in making their choices.

Sen's paradox demonstrates the possibility of conflict between an emphasis solely on self-regarding interests and the inclusion of other-regarding preferences. The problem can be avoided merely by constraining people to have only self-regarding preferences, for cases in which actors have preferences that are other-regarding can be ones in which criteria of choice conflict with one another.[2]

In domestic society the admissability of preferences about others' behavior in arriving at social choices is a matter of philosophical and political debate. It is a matter of public dispute whether society's choices (in the form of laws) should reflect the preferences that individuals have about how others should behave. Some argue that public policy should only reflect peoples' self-regarding preferences. Others argue that all preferences matter, including those that individuals en-

---

2. Indeed this is one way to resolve the impossibility. In the parlance of social choice, it entails dealing with the conditions of unrestricted domain. Among others, see Wulf Gaertner, "Envy-Free Rights Assignments and Self-Oriented Preferences," *Mathematical Social Sciences* 2 (1982): 199–208.

tertain about the desirability or acceptability of others' behavior. Thus, for example, some argue that consenting adults should be allowed to do what they wish in private and not be affected by others' views. Alternatively, people's preferences that others not engage in certain behaviors can be given weight comparable to other preferences.

Although such debates are of interest in discussions of domestic public policy, they are nonsensical in international relations. States do have preferences regarding the behavior of others. And although many inveigh against making relativistic assessments about other countries' behaviors and against intervening in other societies' internal affairs, states are free to make choices by incorporating others' payoffs in their decision calculus and often do so.

The realist position is that actors do have preferences about others and that they make choices using comparative assessments. Indeed the realist underpinnings of the Paretian liberal story lie in the very fact that individuals have preferences about others. People do care about what others are forced to do and are prevented from doing, not just about what they can and cannot do. And in international life there is no entity capable of imposing the collective choice that admits only of nonmeddlesome preferences.

The externalities that lie at the root of relativistic assessments in international politics are also present in domestic politics. States calculate the payoffs to others because these have externalities. They are indeed affected by the payoffs that others experience. The same is true in the Paretian liberal example. Each of the two people is affected by what happens to the other. The prude is affected by the other's reading of the work. The spread of pernicious ideas and practices directly affects a prude's quality of life. The other's quality of life is also affected by the prude's behavior. Neither person feels unconcerned about the other's reading the book. In international relations, states make comparative assessments because they believe they are affected by the payoffs of others.

6

Alliances and Dilemmas of Entanglement

Alliances mark the cooperative end of the continuum of international relations.[1] Unfortunately, scholars have pegged as alliances too wide a variety of cooperative (and even neutral) inter-nation relations. Along with diplomats, they have created an array of terms to describe co-operative relationships, including "alignment," "entente," "détente," "special relationship," as well as "alliance."[2] Alliances themselves have been characterized as "offensive or defensive, limited or unlimited, equal or unequal, bilateral or multilateral," "consultative or automatic, with or without military conventions."[3] In short, we recognize alliances

---

1. For reviews of the literature on alliances, see Ole R. Holsti, P. Terrence Hopmann, and John D. Sullivan, *Unity and Disintegration in International Alliances: Comparative Studies* (New York: John Wiley, 1973), especially chap. 1 and appendix C; Philip M. Burgess and David W. Moore, "Inter-Nation Alliances: An Inventory and Appraisal of Propositions," in *Political Science Annual: An International Review*, vol. 3, ed. James A. Robinson (New York: Bobbs-Merrill, 1972), pp. 339–83; Bruce Bueno de Mesquita and J. David Singer, "Alliances, Capabilities, and War: A Review and Synthesis," in *Political Science Annual: An International Review*, vol. 4, ed. Cornelius P. Cotter (New York: Bobbs-Merrill, 1973), pp. 237–80; and Brian L. Job, "Grins without Cats: In Pursuit of Knowledge of International Alliances," in *Cumulation in International Relations Research*, ed. P. Terrence Hopmann, Dina A. Zinnes, and J. David Singer, *Monograph Series in World Affairs* 18 (1981): 39–63.

2. For an empirical description, see J. David Singer and Melvin Small, "Formal Alliances, 1815–1939: A Quantitative Description," *Journal of Peace Research* 3 (1966): 1–32, and Melvin Small and J. David Singer, "Formal Alliances, 1816–1965: An Extension of the Basic Data," *Journal of Peace Research* 6 (1969): 257–82.

3. Paul W. Schroeder, "Alliances, 1815–1945: Weapons of Power and Tools of Management," in *Historical Dimensions of National Security Problems*, ed. Klaus Knorr (Lawrence: University Press of Kansas, 1976), pp. 227, 255.

to vary in form and content. Notwithstanding the diversity of relationships that can be described as alliances, the concept is central to the
analysis of international politics.

Yet alliances pose a problem for international relations theory. For
both realists and liberals, the international system is anarchic, and the
states that interact within it are autonomous and independent. In this
world any explanation of cooperation must be rooted in the self-interest
of states. Why, then, should alliances exist as formal institutionalized
arrangements? If a certain behavior is in a nation's self-interest, it
should act accordingly and be expected by others to do so. Moreover,
if interests change, so will behavior. We cannot explain why states form
alliances or what difference these arrangements make—why regimes,
alliances, and international organizations should matter.[4] Alliances
should be inconsequential because they do not engender behavior;
rather, behavior and alliances are both products of the underlying pattern of preferences. In the realists' view, states must ultimately rely on
themselves for their own survival. If international politics is the product
of autonomous independent national behavior in an anarchic setting,
there should be no reason for multi-nation groupings to exist. Alliances
should be equally problematic for liberals. Autonomous self-interested
actors engage in mutually advantageous exchange. The international
system, like a market, should reflect discrete interactions and not entail
alliances.[5]

Most international relations theorists, especially realists and balance-
of-power theorists, take the view that alliances represent temporary
marriages of convenience. Indeed, if alliances are to play a role in maintaining balances of power, they must be fluid and flexible, continually
changing as the relative power of states changes. As Morton Kaplan
describes them, alliances lack permanence and are "transient instrumental adjustments to a changing international environment."[6]

4. This is very much like the backdrop for the discussion of regimes in Chapter 2.
5. This is analogous to the problem economists confront in explaining the emergence
of hierarchical firms that internalize certain exchanges in lieu of market transactions.
The development of vertical and horizontal integration, as well as the emergence of the
multinational corporation, is problematic for the classic view of market economics. The
economists' solution is to explain that self-interested actors create hierarchical structures
precisely in order to internalize particular exchanges and so do away with the transaction
costs that sometimes make market exchanges less than optimally efficient. The key figure
associated with this argument is Oliver Williamson, *The Economic Institutions of Capitalism: Firms, Markets, and Relational Contracting* (New York: Free Press, 1985). For an
application of the argument to international relations, see Katja Weber, "Alliances and
Confederations: An Analysis of Cooperative Security Arrangements," Ph.D. dissertation,
University of California, Los Angeles, in progress.
6. Morton A. Kaplan, *System and Process in International Politics* (New York: John

The role of alliances in balance-of-power theory is inherently para-doxical. States concerned with their own survival will act in concert to prevent the emergence of a power that threatens them. This coinci-dence of interests forms the basis for alliances, which in turn undergird a balance of power. Hence these alliances hardly seem necessary. If, on the other hand, alliances entail commitments that states have no in-terest in fulfilling, then nations will not keep to their terms, and the accords will have no consequence. Alliances, then, must be either un-necessary or inherently unbelievable bluffs.[7]

Many alliances reflect nothing more than self-interest. Some alliances have probably been little more than symbolic affirmations of mutual interests that really need not have been institutionalized and formal-ized. Alternatively, some alliances have really been coordination or col-laboration regimes, forms of institutionalized cooperation that entail joint decision making to resolve dilemmas that arise from individu-alistic decision making. Both symbolic alliances and those that are nothing more than coordination and collaboration regimes reflect con-stellations of individualistic self-interest.

Yet some alliances, more constraining than regimes, affect interests as well as mirror them. Indeed, some theoretical arguments recognize alliances as arrangements that bind and restrain the exercise of self-interest. In a classic article linking multipolarity with systemic stability, Karl Deutsch and J. David Singer build their argument on the critical assumption that alliances reduce the effective number of independent actors in the international system.[8] Or in the words of Michael Altfeld,

---

Wiley, 1957), p. 116; also see preface, p. 29. In a recent essay one scholar argues that the very security dilemma that exists between potential and actual enemies exists within alliances; see Glenn H. Snyder, "The Security Dilemma in Alliance Politics," *World Politics* 36 (July 1984): 461–95.

7. Alliances might be inconsequential for the behavior of allies but still fulfill a useful role as devices to signal to third parties intentions about contingent future behavior that might not otherwise be presumptive. States signal commitments to their allies to deter others from attacking these allies. For the determinants of successful extended deter-rence, see Bruce M. Russett, "The Calculus of Deterrence," *Journal of Conflict Resolution* 7 (June 1963): 97–109; Clinton F. Fink, "More Calculations about Deterrence," *Journal of Conflict Resolution* 9 (March 1965): 54–65; and Paul Huth and Bruce Russett, "What Makes Deterrence Work? Cases from 1900 to 1980," *World Politics* 36 (July 1984): 496–526. Such signaling is important because these interests and behaviors might not otherwise be assumed by potential aggressors. One implication of this reason for alliance formation is that without alliances, other states might draw different inferences about the interests and behaviors of the parties. In other words, alliances signal the existence of interests different from those that might otherwise be observed.

8. Karl W. Deutsch and J. David Singer, "Multipolar Power Systems and International Stability," *World Politics* 16 (April 1964): 390–406.

"the cost of an alliance to a government is computed in terms of...
autonomy."[9]

In this chapter I delineate a different argument: some alliances do
matter because they lead the states bound by them sometimes to pursue certain courses of action because of the needs of their allies and
in contradiction to their own self-interest. The argument made here,
both analytically and illustratively, is that some alliances are more than
collaborative regimes—they are arrangements for joint decision making
in which states care about the joint power of the alliance and in which
they attach weights to their allies' interests. They commit themselves
to act neither unilaterally nor without concern for the needs of their
allies. Such alliances entail collective interests, sympathy, and soli-
darity.[10]

Nations form alliances to deal with deficiencies of power. Self-interested and autonomous states join to offset their relative weaknesses
vis-à-vis stated and potential enemies. Because the power of the coalition exceeds that of individual states, its creation aids each in its
search for security.[11]

One immediate implication is that states value the military power of
their allies. In stark contrast to the point made in the previous chapter,
that rivals fear one another's military power, allies view the armed might
of their partners with neither indifference nor fear. They see this additional military strength as an increment to their own. Indeed, the
aggregation of capabilities is one defining characteristic of alliances.[12]
As some scholars argue, states at times seek allies rather than procure
more weapons unilaterally. Hence they may actually be able to reduce
their defense spending as their allies spend more.[13] Such an under-

9. Michael F. Altfeld, "The Decision to Ally: A Theory and Test," *Western Political
Quarterly* 37 (December 1984): 526. Altfeld argues that alliances tie nations to their allies'
positions, and he cites scholars who have made that or similar assumptions. Also see
propositions V3 and V17 in Holsti, Hopmann, and Sullivan, *Unity and Disintegration in
International Relations*, pp. 276–77.

10. Throughout this chapter, it is important to remember that not all alliances are
characterized by the dynamics described here.

11. Although alliances are formed because of considerations of power and the aggregation of military capabilities, states also enter alliances to constrain and manage their
allies. For an essay that stresses the role of alliances as tools of control and management,
see Schroeder, "Alliances, 1815–1945."

12. See, for example, Julian R. Friedman, "Alliance in International Politics," in *Alliance
in International Politics*, ed. Julian R. Friedman, Christopher Bladen, and Steven Rosen
(Boston: Allyn and Bacon, 1970), p. 5.

13. Altfeld, "The Decision to Ally," and the discussion of negative reaction functions
in Philip A. Schrodt, "Richardson's N-Nation Model and the Balance of Power," *American
Journal of Political Science* 22 (May 1978): 364–90. The application of collective-goods

standing of states as prepared to rely on others, even to a minimal degree, and to reduce their own defense efforts, undercuts the view of international politics as a system of self-help.

The difference in how states perceive allies' and rivals' arsenals is readily apparent in the way that American policymakers view the deployment of nuclear weapons. Although they see any new Soviet deployment as a threat they must counter, they consider additions to the British nuclear arsenal as strengthening the West against the Soviet challenge.

Anglo-American Nuclear Relations and Polaris

U.S. policy toward Britain since the Second World War provides an excellent example of a nation paying attention to the needs of its ally as well as to its own concerns. It also illuminates the impact of nuclear weapons on an alliance that predated their existence yet survived as a partnership despite the independent development of such arms by both nations. The tension between nuclear collaboration and autonomy did, however, lead to one of the alliance's greatest crises, as well as to the relationship's strong reaffirmation in the resolution of that crisis. The Anglo-American dispute over the U.S. cancellation of its plans to develop the Skybolt missile, as well as its subsequent resolution in the American commitment to provide Polaris missiles to the British in Skybolt's stead, demonstrates the conflict that existed for the United States between pursuing individualistic and joint interests.

The United States and Great Britain, not simply allies when the crisis erupted in December 1962, had what both nations termed a "special relationship." American-British ties were so strong that many contemporary observers saw Great Britain more as an Atlantic nation linked to the United States than a European nation tied to the continent.[14] The United States and Britain had fought together in both world wars and were NATO allies. They had collaborated on harnessing atomic energy during the Second World War.

The American desire for nuclear monopolism led to the brief collapse

---

arguments about free riders to alliances also carries that implication; see Mancur Olson, Jr., and Richard Zeckhauser, "An Economic Theory of Alliances," *Review of Economics and Statistics* 48 (August 1966): 266–79.

14. Indeed many believe de Gaulle's veto of British entry into the European Economic Community to be directly related to the reaffirmation of the American-British special relationship symbolized in the resolution of the Skybolt crisis.

of this partnership in the late 1940s and early 1950s.[15] During this period
Great Britain developed its own nuclear weapons and was committed
to maintaining an independent nuclear force, relying on bombers to
carry those weapons. By the end of the 1950s, however, it had become
clear that the aging British bomber fleet would become obsolete and
would have to be replaced by some other delivery system.

In 1960 President Eisenhower promised British Prime Minister Macmillan that the United States would make the Skybolt missile available
to Britain. This air-to-ground missile would maintain the utility of the
planes, since they would no longer have to penetrate Soviet air space
in order to bomb enemy soil. Although the president's pledge was contingent on the Americans' actually proceeding to develop Skybolt, the
British nevertheless scrapped their attempt to develop their own missile (Blue Streak). At the time of Eisenhower's promise, Britain made
available to the United States facilities at which U.S. Navy Polaris submarines could be based. Although there existed no formal link between
the promise of Skybolt and the offer of port facilities, the British considered the United States to be under a "moral obligation" to provide
the strategic missile.[16]

In 1961 a new administration came to office in the United States,
committed both to rationalizing American defense procurements and
to restructuring American-European relations. Given the successful development of both American land-based and sea-based missile forces,
Minuteman and Polaris, and the cancellation of plans for a new American strategic bomber, the Kennedy administration had no interest in
pursuing the Skybolt project. The new missile also failed its initial tests.
The Kennedy administration, without adequately calculating the diplomatic fallout, proceeded to cancel the program altogether.

By the time of the Skybolt affair in late 1962, the new American president had already established a very good relationship with Great Britain. He and Prime Minister Macmillan had met four times in 1961
alone.[17] The British ambassador to the United States, David Ormsby-

15. On the immediate postwar period, see Andrew J. Pierre, *Nuclear Politics: The
British Experience with an Independent Strategic Force, 1939–1970* (London: Oxford University Press, 1970), pp. 112–20, 130–33, and John Baylis, *Anglo-American Defense Relations, 1939–1984*, 2d ed. (New York: St. Martin's Press, 1984), pp. 30–34, 41–45. Also see the
essays by Bradford Perkins, D. Cameron Watt, and Alistair Horne in Wm. Roger Louis and
Hedley Bull, eds., *The 'Special Relationship': Anglo-American Relations since 1945* (Oxford:
Clarendon Press, 1986).
16. Pierre, *Nuclear Politics*, p. 227.
17. Macmillan met with Kennedy seven times, more than any other Western leader,
during the latter's short time as president.

Gore, was a longtime friend of the president's, a man of whom Kennedy once said, "I trust David as I would my own Cabinet."[18]

Yet the announcement of Skybolt's cancellation led to a test of the American-British relationship. At a summit meeting at Nassau, a conference scheduled before the crisis began, Macmillan emphasized the history of Anglo-American nuclear collaboration to demonstrate that the new U.S. position represented a major break with past practice and current expectations. In addition, Macmillan detailed his own precarious political position at home, for he and the Conservative party had promised to maintain an independent nuclear force with American help. Now he had become open to attack not only from the left, which did not want an independent force, but from the right for having chosen to rely on another nation rather than going it alone. He was, he said, "like a ship that looked buoyant but was apt to sink."[19]

The president was determined to give Macmillan something, but the prime minister had only one request. Refusing every alternative Kennedy offered, he insisted that he would only accept Polaris missiles, which could be fired from British submarines and outfitted with British nuclear warheads. This would give Great Britain the equivalent of the advanced sea-based strategic component of America's own nuclear arsenal.

The British desire for Polaris posed great problems for the United States, which did not want to comply with Macmillan's request. Firmly opposed to nuclear proliferation, the Kennedy administration wanted to restrain the development of independent limited nuclear forces. In his famous address in Ann Arbor, Michigan, the speech in which he unveiled the doctrine of flexible response, Secretary of Defense McNamara described such independent forces as "dangerous, expensive, prone to obsolescence and lacking in credibility as a deterrent."[20] He restated this view at the NATO Ministerial Conference only a week before the Nassau summit.[21] The administration was also committed to a doctrine of a controlled and flexible response to aggression in Europe, a policy which would entail either unilateral American deci-

18. Theodore C. Sorensen, *Kennedy* (New York: Harper and Row, 1965), p. 559. One chapter in David Nunnerley's *President Kennedy and Britain* (New York: St. Martin's Press, 1972), devoted to Ormsby-Gore, is entitled "A Special Relationship within the 'Special Relationship.'"

19. Pierre, *Nuclear Politics*, p. 234.

20. Quoted in ibid., pp. 208–9, among others.

21. Criticism of the British deterrent had been voiced by many members of the administration including President Kennedy. See Nunnerley, *President Kennedy and Britain*, p. 122.

sions or joint decisions in a broader body such as NATO. To give Britain
Polaris would upset those European nations not developing nuclear
forces of their own, and it would prevent the United States from controlling the nature and course of escalation in Western Europe.

Given the new American administration's general views about independent nuclear forces and proliferation, the British wondered whether
the cancellation of Skybolt was specifically aimed at them rather than
a decision made for other reasons. When McNamara officially informed
British Defense Minister Thorneycroft of the cancellation, Thorneycroft
asked if the United States "wanted to deprive Britain of its 'independent
deterrent role.'"[22] At the very least, the cancellation represented, in the
words of British Prime Minister Macmillan, "evidence of American dis-
regard for our interests."[23] President Kennedy assured Macmillan at
Nassau that the decision to cancel Skybolt development was made on
technical grounds and not for political reasons.[24] Macmillan accepted
these assurances and understood the Americans to be "determined to
kill Skybolt on good general grounds—not merely to annoy us or to
drive Great Britain out of the nuclear business."[25] Indeed, the United
States offered to revive Skybolt and share its development costs with
the British. But this not only would cost the British more than they
had originally assumed but would require their accepting a strategic
option whose technical merits the Americans had already publicly
maligned.

But reassuring the British about the reasons for Skybolt's cancellation
and offering to revive it did not translate into a desire to give them
Polaris. At Nassau Kennedy's advisers opposed giving Polaris to the
British, and some of them even saw the situation as an opportunity to
end the preferential treatment given to Britain.[26]

Macmillan readily admitted that the British had been given only a
"firm, although not legal, assurance" that they could get Polaris if Skybolt were not to be developed.[27] Indeed it had merely been a gentle-
men's agreement. In fact, although the British had offered Polaris basing
rights "more or less in return for Skybolt," they conceded that there

22. Pierre, *Nuclear Politics*, p. 230.
23. Harold Macmillan, *At the End of the Day, 1961–1963* (New York: Harper and Row,
1973), p. 344.
24. Ibid., p. 357.
25. Ibid., p. 361.
26. See Arthur M. Schlesinger, Jr., *A Thousand Days: John F. Kennedy in the White
House* (Boston: Houghton Mifflin, 1965), p. 860; Sorensen, *Kennedy*, pp. 566–67; and Baylis, *Anglo-American Defense Relations*, pp. 101, 103.
27. Macmillan, *At the End of the Day*, p. 342.

was no direct linkage between the two.[28] Further, Macmillan was prepared, absent an agreement on continued nuclear cooperation, to part with "honour and dignity" and said that he would not break the commitments he had made regarding Polaris bases even if the United States did not live up to its part of the bargain.[29] The American delegation at Nassau insisted that it would not provide Polaris and, Macmillan recounted later, "continued to maintain [in an often heated discussion] that the change from Skybolt to Polaris was one of principle to which they were not even honourably committed."[30]

In the end, the United States agreed to provide Polaris missiles to the British. President Kennedy believed that the special relationship with the British required him to provide an alternative to President Eisenhower's promise of Skybolt. As Kennedy later said to Theodore Sorensen, "Looking at it from their point of view . . . it might well be concluded that . . . we had an obligation to provide an alternative."[31] For some members of the American delegation, the decision to provide Polaris to the British "was a missed opportunity and bitter defeat."[32] After the summit, the president responded to comments that he was soft on Macmillan by noting the prime minister's domestic political difficulties in the wake of Skybolt's cancellation. "If you were in that kind of trouble," Kennedy said, "you would want a friend."[33] His conviction comported with that of President Eisenhower, who had told American officials attempting to work out nuclear cooperation with Britain not to "be too lawyer-like. A great alliance requires, above all, faith and trust on both sides."[34]

This historical episode in Anglo-American relations illustrates a number of points. First, the United States acted in light of its ally's needs as well as its own. It was not in the interest of the United States either to develop Skybolt for the British or to provide Polaris to them. After all, the United States had decided to abandon its plans to develop Skybolt simply in terms of its own needs and without giving thought to British interests. When the American delegation offered at Nassau to cover half the costs of developing the missile, it did so only to assist the British.

---

28. Quoted in Baylis, *Anglo-American Defense Relations*, p. 99.
29. Pierre, *Nuclear Politics*, p. 235.
30. Macmillan, *At the End of the Day*, p. 358.
31. Sorensen, *Kennedy*, p. 566. Also see p. 567 and Schlesinger, *A Thousand Days*, p. 862.
32. Schlesinger, *A Thousand Days*, p. 865.
33. Sorensen, *Kennedy*, p. 559.
34. Eisenhower, *Waging Peace, 1956–1961*, p. 219, quoted in Pierre, *Nuclear Politics*, p. 143.

Further, the United States agreed to give Britain Polaris even though it did not want to do so. Since doing so undercut American policy, Kennedy tried hard to avoid providing it and looked for an alternative the British might accept. Second, the United States acted in the interests of its ally when failing to do so would not have entailed recognizable costs. The British did not threaten to renege on any of their commitments to the United States. They would not prevent the United States from basing its submarines in Britain. There was no possibility that the alliance might rupture. Third, the American decision was made without knowledge of the ultimate consequences, either for Macmillan or for American-British relations, of sending the prime minister home empty-handed. In short, the Americans clearly perceived there to be a special relationship, one that imposed requirements on both nations to maintain it. The United States knew what Britain wanted. Although British desires would involve a cost to the United States, President Kennedy placed some weight on Macmillan's needs and interests and not just on those of his own country.

As a result of this decision, the British independent nuclear force has been maintained with American assistance, relying since the middle 1960s, on first the Polaris and then the Trident missile system. Ironically, although the British want to maintain an "independent" nuclear force, they remain dependent on the United States to help them do so.[35]

It is no simple matter to explain this American decision to supply nuclear weapons delivery systems to Great Britain. After all, the United States was committed to providing a nuclear umbrella to ensure the defense of Western Europe and Japan. Hence no clear need existed for the British to maintain an independent system. The United States was on record as opposing nuclear proliferation, and the secretary of defense had publicly and privately derided independent limited nuclear forces. From the U.S. perspective, it would certainly have been preferable to have the British pay the Americans to deploy additional submarines with nuclear missiles rather than build them for sale to its ally.[36] Many options were preferable to providing missiles that would

35. The British desire for an independent force is driven in part by their concern about the depth of the American commitment to deter aggression and defend Western Europe, and ironically, the United States deals with this concern, in part, by giving the British weapons that could also be used to destroy American cities.

36. Indeed, the Kennedy administration pressed the Germans to help pay for American forces in Germany.

not be under American control and which could then be used as the British wished, even against U.S. interests and against American soil.[37]

In short, the decision to provide Polaris to Great Britain cannot be explained by reference to American self-interest. A unilateral calculation of such interest clearly entailed a first preference for providing Britain with nothing, and then giving Britain a variety of alternatives short of Polaris. Nor can the American decision be explained as the costly price of retaining basing rights. At no point did the British make any threats. To the contrary, they reaffirmed British commitments.

The United States made its calculated decision to provide Polaris by attaching some weight to British needs and concerns. Kennedy knew the value that Macmillan placed on Polaris. Indeed, Macmillan attached more weight to getting the missile than the United States attached to not providing it. In effect, President Kennedy maximized joint returns rather than American ones. Only such a decision criterion, assessing alliance interests rather than American interests narrowly conceived, captures the basis of American calculations. In fact, the debate within the American delegation was about this very question of whether to attach weight to British concerns. Those opposed to giving the British Polaris stressed American interests and deemphasized the special relationship between the two nations and any obligations it might entail. Others focused on the importance of meeting British needs and concerns.[38]

That Anglo-American relations cannot be explained merely by individualistic self-interest is recognized by scholars who have tried to describe and analyze it. G. M. Dillon uses terms such as "civility," "friendship," "trust," and "intimacy" to explain the extent to which the special relationship cannot be explained merely by interest.[39] Raymond Dawson and Richard Rosecrance argue that the history of the Anglo-American partnership "demonstrates that an alliance may itself gen-

37. It should be noted that the maintenance of an alliance between two nuclear powers, even without such military assistance, already represents a falsification of the widely stated proposition that possession of nuclear weapons by alliance members leads to the disintegration of the alliance. See Holsti, Hopmann, and Sullivan, *Unity and Disintegration in International Alliances*, pp. 25–28, and the propositions in appendix C referenced there.

38. No one argued that Great Britain was a rival that should not be armed or assisted in any fashion. There was never a question of maximizing relative gains vis-à-vis the British.

39. G. M. Dillon, *Dependence and Deterrence: Success and Civility in the Anglo-American Special Nuclear Relationship, 1962–1982* (Aldershot, England: Gower Publishing, 1983), chap. 1, and especially p. 13.

erate a set of interests that become central to the formulation of policy by the members of the system, and that the preservation of the alliance may be a salient national objective, overriding egocentric calculations of interest." Indeed, they propose the concept of "alliance interests" to explain national behavior.[40] My own formulation is that in such alliances, nations sometimes attach some weight to the needs and concerns of their allies and choose to maximize the joint interests of the alliance rather than their own national self-interest.

Most important, the Kennedy administration did not see increments of British strategic power as threatening to the United States. Although such forces posed a political problem and might complicate American strategic calculations in a crisis, they also represented additions to Western deterrent power against the Soviets. The United States did not see a nuclear delivery system in the hands of the British as a danger, even though its missiles could be as easily targeted on Washington as on Moscow.

Just as states are not indifferent to the arms procured by their rivals, so they are not indifferent to those procured by their allies. But whereas a rival's military forces are viewed as a threat, an ally's military forces provide reassurance. Whereas increases in a rival's military power degrade the efficacy of one's own arsenal, deployments by an ally augment one's own military forces. Thus, although a rival's deployments entail negative externalities, those of an ally can have *positive externalities*. The military utility of an alliance derives from the combined impact of its members' military forces.

Many arms transfers are to allies and involve weapons of such sophistication as to undercut the supplier's military advantage should the relationship ever change. Such situations are impossible to explain if one assumes that states always act to maximize their relative power with every other nation. Such situations remain difficult to explain on the grounds of individualistic self-interest.[41]

40. Raymond Dawson and Richard Rosecrance, "Theory and Reality in the Anglo-American Alliance," *World Politics* 19 (October 1966): 22. They make the point not only for the United States but also for Britain. Moreover, virtually their entire discussion of alliance interests in the Anglo-American relationship is about the period before December 1962. Finally, they attack conventional alliance theory, which, they argue, cannot explain why a world power would "endow lesser states with the attributes of strategic independence" (pp. 50–51).

41. Arms transfers can reflect malevolence and individualism as well as an interest in building the power of an ally. Some arms transfers represent mere exchanges of weapons for money. The weapons sold are seen simply as a commodity exchanged for money rather than as decrements or increments to national security. Indeed the payment serves as a claim on future output, whereas the weapons, in and of themselves, merely eat

An interesting illustration of how arms transfers are viewed as a function of one state's relationship with another is provided by the evolution of American views of Iran. In the 1970s the president of the United States gave an unprecedented order instructing the national security bureaucracy to provide the Shah of Iran with whatever weapons he requested.[42] By 1987 Iran was ruled by a government hostile to the United States, and the Reagan administration was damaged by revelations that it had shipped obsolete weapons to the Ayatollah. In Congressional hearings, Congressman Aspin asked Secretary of Defense Weinberger whether the Pentagon had assessed the impact of the arms shipments on any future military conflict between Iran and the United States. The geopolitical situation had not changed, and there was widespread hope of a future American-Iranian relationship based on opposition to the Soviet Union. But the interests and orientation of the Iranian government had shifted, and the perception and evaluation of weapons sales had altered dramatically, from a view that they would be used in line with American interests to the possibility that they might be used against American forces.

In short, states view the arsenals of other states through the prism of their relationship. The weapons of rivals are threatening, and states go to great lengths to prevent arms transfers to their enemies. The arms of an ally are seen in a fundamentally different light.

Alliance Dilemmas

Alliance dilemmas, or what I would also call dilemmas of entanglement, arise when individualistic self-interest points in a different direction from conjoint interest. This situation after all, is the true test of an alliance. When individualistic and joint interests converge, there is no need to posit the existence of joint interests, and hence there is no reason for alliances to exist. When individualistic and conjoint interests diverge, actors must choose between their own egoistic concerns and the interests of the joint entity, the alliance. If actors always choose their individualistic interests when such situations arise, alliances are

---

capital rather than generate it. In some cases arms may be transferred to harm the recipient and so reflect malevolence and competition. As Harry Truman said when he was a senator in June 1941, "If we see that Germany is winning we ought to help Russia and if Russia is winning we ought to help Germany and that way let them kill as many as possible"; *New York Times*, June 24, 1941, p. 7. Divide-and-conquer strategies often entail seemingly beneficent assistance to serve malevolent objectives.

42. See Gary Sick, *All Fall Down: America's Tragic Encounter with Iran* (New York: Random House, 1985), pp. 13–21.

neither efficacious nor meaningful. Truly meaningful alliances are those in which that dilemma is, at least sometimes, resolved in favor of conjoint interests.

Only when the consequences of a state's choices have more of an effect on others than on itself will it be torn between a dominant strategy of maximizing its own interest and one of maximizing the interest of the group (see Appendix 1). It is such an asymmetry of impact that is at the root of dilemmas of entanglement.

Ironically, alliance dilemmas arise in the presence of the same kind of asymmetry that gives rise to dilemmas of competition. Both dilemmas emerge when a state's decision has a greater impact on others' payoffs than on its own. The dilemma of rivalry entails a choice between a state's absolute and relative payoffs. An alliance dilemma involves a state's choice between absolute and joint payoffs when the difference in its absolute payoff is dwarfed by the payoff difference for its ally.

States aid their allies even at their own expense because it is in their interest to do so.[43] Sustaining alliance relationships is a reason for maximizing joint interests rather than narrow national self-interest. Seemingly self-abnegating behavior can be seen as reflecting merely long-run self-interest. But such behavior is not based on incorporating one's future payoffs into current calculations. To reduce present self-abnegation to the incorporation of future payoffs presumes that actors have some sense of the future payoffs that will accrue from short-term self-abnegation. In international politics, as in interpersonal relations, actors rarely know (or can even estimate) future payoffs. Rather, actors know the opportunity and actual costs of self-abnegation, and they also know the payoffs to others of such a strategy. In situations in which different strategies maximize joint and individual interests, states must decide whether individual self-extension or conjoint self-abnegation is in their best interest. This becomes the benchmark for assessing the value of the relationship. What a state is willing to do for its allies in such circumstances is a measure of the relationship.[44] Thus seemingly nonself-interested behavior can be described as self-interested by attaching some value to the alliance. Still, such actions have been chosen using a decision criterion that maximizes joint rather than individual interests.[45]

43. See Appendix 2 on altruism and self-interest.
44. Or it is a measure of the value it attaches to its reputation. If nonself-interested behavior maintains one's reputation, and if the value attached to reputation exceeds the payoff forgone, a self-interested explanation is available for seemingly self-abnegating choices.
45. To repeat, I am not interested in whether maximizing joint interests is character-

This is evident in the American decision to provide the British with Polaris. That decision cannot be explained by saying that the United States considered discounted future returns and thus maximized its long-term self-interest. Clearly, a decision to retain the special relationship was a decision that recognized benefits to be obtained from maintaining the partnership. But the American officials making the decision in December 1962 did not have a concrete sense of what future American returns from the relationship might be. Nor did they know how adversely affected the United States might be should they send Macmillan home without Polaris. In fact, they would most likely not jeopardize the special relationship even if they did deny Macmillan's request. What they did know was the value that Macmillan placed on Polaris. He attached more weight to having Polaris than the United States attached to not providing Great Britain with the missile.

States are willing to bear costs to sustain relationships for a reason. Alliances are embedded in a larger environment. In an anarchic international system of more than two nations, no state can pursue relativistic strategies in all of its bilateral relations.[46] Minimally, states compete with rivals and with enemies but maximize absolute egoistic self-interest in bilateral relations with nonenemies. In a world with more than two actors, rivalries, threats, and insecurity make alliances necessary.[47] Through alliances, conjoint interests emerge in an international environment of conflict and hostility, a world rooted in competition and individualism.[48] American-British relations following World War II, for example, cannot be understood without reference to the Cold War with the Soviet Union.

---

ized as nonself-interested behavior or made self-interested by inclusion of a value for the relationship. I want to emphasize two points. First, actors do not maximize discounted present value because they do not know the magnitudes of a future payoff stream. Second, actors make their choices by assessing the values their allies attach to their self-abnegating strategy.

46. The need to distinguish "archrivals" from others, I would argue, is rooted in the typical inability simultaneously to pursue relative gains vis-à-vis multiple rivals. The net result is a relativistic focus on a particular archrival combined with the pursuit of individualistic self-interest in bilateral relations with others.

47. Game theory is typically divided into noncooperative and cooperative games. The latter are distinguished by the existence of mechanisms to enforce agreements. When there are more than two actors, cooperative games are used to study coalitions. The assumption of enforceable agreements, however, is problematic for international relations.

48. This argument is related to the one that holds that conflict with an external group increases cohesion among group members. For a review of that literature, see Arthur A. Stein, "Conflict and Cohesion: A Review of the Literature," *Journal of Conflict Resolution* 20 (March 1976): 143–72.

The nature of disagreement among allies differs from other international conflicts of interest. Conflicts of interest can emerge and destroy alliances. The constellation of states' interests can shift, and their relationships can follow suit. But there are also conflicts of interest that uniquely emerge in strong alliances.

Alliances, like marriages, are continually under stress. Issues constantly arise that put the alliance to the test, that determine whether the ties continue to bind and that set the price the allies are willing to pay to sustain the relationship. Conflicts within an alliance differ, in part, because they develop against a backdrop of past cooperation. Indeed some conflicts among allies presume past intimacy; they can arise only between states that are strongly linked. A trade war, for example, can occur only among states that trade extensively with one another.[49] Trade wars arise when a nation uses its economic interdependence with another state to improve the relative terms of exchange, when one country changes its trade laws to improve its position vis-à-vis its most important trading partners. Trade wars do not break out between enemies who have few links with one another. The United States, for example, embargoes goods that can be exported to its enemy, the Soviet Union, but is involved in trade disputes only with its closest allies, those with whom it has the most developed trading relations.

Similarly, the desire to renegotiate the terms of a relationship is often the root of other forms of conflict between allies pursuing conjoint interests. A desire to shift the burden, or change the weights that each attaches to the other's interests, becomes the basis for disputes. The irony is that such conflicts can develop only between states that have close ties.[50]

International disputes between conjoined allies are not unlike quarrels between couples. Many of the fights between a husband and wife could not occur if they were not married. People who have not been intimate cannot fight about how to raise their children. Yet the mere

49. This example is drawn from Arthur A. Stein, "Governments, Economic Interdependence, and International Cooperation," in *Behavior, Society, and Nuclear War*, vol. 3, ed. Philip E. Tetlock, Jo L. Husbands, Robert Jervis, Paul C. Stern, and Charles Tilly (New York: Oxford University Press, for the National Research Council of the National Academy of Sciences, forthcoming). Also see Stein, "The Hegemon's Dilemma: Great Britain, the United States, and the International Economic Order," *International Organization* 38 (Spring 1984): 355–86.

50. This is also the root of a phenomenon often observed in American foreign policy, that the United States sometimes seems to expect more from its allies and leans more on its allies than on its enemies. The retort to such criticisms is that the United States has leverage in such relationships. The leverage, of course, is its ability to threaten the relationship unless American interests are given preeminent weight.

existence of a relationship does not ensure that there will be amicable resolution of disagreements and disputes. Just as couples get divorced over issues that arise in the wake of their marriage, so do international alliances dissolve.[51]

Misperceptions and Alliances

Misperception can matter even in an alliance in which the actors maximize conjoint interests. In such a partnership, a state makes its decision partly in light of its assessment of the interests and needs of its ally. To misperceive the ally's interests either does not matter or leads to conflict.

A conflict-generating misperception is unintended when one state wrongly believes that its ally has no interest in how it makes a particular decision. Something of this sort happened in the case of the American cancellation of Skybolt. The American decision was made on technical grounds without attention to British interests and concerns. The Americans might have made the same decision even if they had taken British concerns into account, but they would presumably have developed a plan to deal with the international political fallout and so have saved both the United States and Great Britain the trauma of the ensuing crisis. A benign misperception would not matter if a state's self-interest led it unknowingly to opt for the strategy that was in its ally's interest. On the other hand, when individualistic interests diverge, a misperception of an ally's interest generates conflict and hostility.

There also exists, however, a possibility for motivated differences in perception. States often act paternalistically by considering their ally's needs but having a different view of them than does the ally itself. This leads to a perverse conflict between nations, in which a state argues that it knows where its ally's best interest lies. Not surprisingly, a state subjected to such paternalistic behavior does not believe its ally's explanation for its actions and assumes the ally is acting according to its own, divergent, individualistically calculated interests. As in conflicts between parents and children, paternalistic decisions are the cause of

51. It should be noted that even attaching weight to the interests of others is no guarantee that problems of strategic interaction will be resolved. Indeed, as O. Henry's "Gift of the Magi" illustrates, even mutual altruism need not resolve incongruent interests. Formal analyses also generate this conclusion; see Norman Frohlich, "Self-interest or Altruism: What Difference?" *Journal of Conflict Resolution* 18 (1974): 55–73; Bruce D. Fitzgerald, "Self-Interest or Altruism: Corrections and Extensions," *Journal of Conflict Resolution* 19 (September 1975): 462–79; and Frohlich, "Comments in Reply," *Journal of Conflict Resolution* 19 (September 1975): 480–83.

disputes between allies. Some of the disputes between the United States and other members of NATO and between the United States and Israel are rooted in paternalistic calculations.

Conclusion

Alliances are unnecessary when there is a congruence of interest. There exists no need for them when independent decision making results in mutually desirable outcomes. Chapter 2 details the argument that there are situations in which self-interested individualists create institutions to ensure collaboration and coordination. That chapter ends with speculation about the longevity of such regimes and whether such regimes outlive changes in underlying interests. The argument here is that there are institutional arrangements that represent still greater levels of cooperation, alliances in which the states attach some weight to an ally's interests and in which states maximize conjoint interests.

Such alliances are not unlike the institution of marriage. Marriage represents the commitment of individuals to each other, more than a temporary and convenient convergence of interests. It signals a commitment to attach some weight to another person's needs and dislikes. It is a commitment to some form of joint decision making. Like alliances, marriages are not forever. They can end in divorce. Nevertheless, marriages are difficult to dissolve. Their dissolution and consummation are consequential. So it is with some alliances, which involve more than the mere convergence of interest. Although they are not permanent, they are not dissolved even following changes in the partners' interests.

Alliances are important institutions between the poles of convergent interests and self-interested autonomy. At one end of the spectrum, alliances do not matter; states act autonomously and in their own individualistic self-interest. At the other end of the spectrum are those situations in which states find that their interests converge. In these cases alliances are irrelevant and stipulate courses of action that would be pursued autonomously.[52] At these two extremes, alliances either do not exist or do not matter.

Alliances matter when states choose conjoint interests when con-

---

52. Moreover, there is no need to join an alliance and constrain oneself if one can obtain the benefits without any of the costs. When one cannot be excluded from sharing the benefits, convergent interests will lead one to become a free rider rather than an ally. The French have all the benefits of belonging to NATO with none of the constraints to their freedom of action or other costs of membership.

fronted by a conflict between individualistic self-interest and joint interests. Such relationships are rooted in a larger environment of competition and conflict. They are sustained by states prepared to pay some absolute individual costs in order to ensure the maintenance of the relationship.

Appendix 1
Implication of a Different Dominant Strategy for Different Decision Criteria

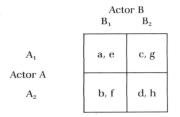

If A_2 is a dominant strategy for absolute returns, then

$$b > a \quad \text{and} \quad d > c$$

If A_1 is a dominant strategy for joint gains, then

$$a + e > b + f \quad \text{and} \quad c + g > d + h$$

By algebraic manipulation:

$$(e + g) - (f + h) > (b + d) - (a + c)$$

By definition:
A's reflexive control over its own returns: $(b + d)/2 - (a + c)/2$
A's fate control over B: $(e + g)/2 - (f + h)/2$

NOTE: Reflexive control and fate control are from Harold H. Kelley and John W. Thibaut, *Interpersonal Relations: A Theory of Interdependence* (New York: John Wiley, 1978), pp. 31–43.

Appendix 2
Altruism and Self-Interest

Self-abnegating behavior is evident among animals and people as well as nations. There are literatures in each of the social sciences, as well as in biology and philosophy, devoted to the study of altruism.[1] There is great disagreement, however, about the nature of altruism and whether there is a self-interested basis to altruistic behavior.

There are a variety of definitions of "altruism." Some define it as behavior intended to maximize the payoffs of others without regard to oneself. Others provide a stronger definition that requires that an actor actually sacrifice and suffer some loss in assisting others in order to be considered an altruist. At the other extreme, "altruism" is defined as an actor's making any choice that involves attaching some weight to the payoffs of others. In this view, altruism is merely one actor's inclusion of another's utility in its own utility function.[2]

A willingness to aid others can be global, or it can be conditional. Some will help all others; some will not. Some will help only those less well off. Some will give only if they are well off. Such giving is conditioned on the ex ante circumstances of donor and recipient. In the case of charity, some people have a generalized desire to give and do so regardless of their income, whereas others give only if they are wealthy. Some are prepared to transfer wealth regardless of the financial circumstances of donor and recipient; others are generous only when they are rich, and give to the poor.[3]

1. As examples, see David Messick and Charles G. McClintock, "Motivational Bases of Choice in Experimental Games," *Journal of Experimental Social Psychology* 4 (January 1968): 1–25; Norman Frohlich and Joe Oppenheimer, "Beyond Economic Man: Altruism, Egalitarianism, and Difference Maximizing," *Journal of Conflict Resolution* 28 (March 1984): 3–24; Thomas C. Schelling, "Altruism, Meanness, and Other Potentially Strategic Behaviors," *American Economic Review* 68 (May 1978): 229–30; and Amartya K. Sen, "Rational Fools: A Critique of the Behavioral Foundations of Economic Theory," *Philosophy and Public Affairs* 6 (Summer 1977): 317–44. Also see other work cited below as well as publications cited by these authors.

2. This has a long-standing tradition in classical economics; see George Stigler, "The Development of Utility Theory," *Journal of Political Economy* 58 (1950): 307–24, 373–436. Also see Thomas Wilson, "Sympathy and Self-Interest," in *The Market and the State: Essays in Honour of Adam Smith*, ed. Thomas Wilson and Andrew S. Skinner (Oxford: Oxford University Press, 1976), pp. 73–112.

3. Harold M. Hochman and Shmuel Nitzan, "Concepts of Extended Preference," *Journal of Economic Behavior and Organization* 6 (1985): 161–76, distinguish between altruism, sympathy, generosity, benevolence, and charitability. In manifesting all of these, they argue, individuals attach utility to others' payoffs. The situations are distinguished by the rates of substitution (as between retaining and transferring income) at different ex ante income allocations and given different rates of exchange. It is also possible that individ-

Altruistic behavior is problematic for any system of explanation rooted in individualistic self-interest. Economists, for example, have built an entire theoretical edifice around individualistic egoism, and yet altruism entails attaching weight to the interests of others.[4] It also poses a problem for evolutionary biology and any system of explanation based on natural selection which eliminates behavior that reduces individual fitness. After all, aiding others rather than maximizing individual fitness reduces one's own ability to survive.

The analytic problem is most readily resolved by reducing altruism to egoistic individualism. Thus self-abnegation can be explained by the expectation of future payoffs. Individuals are merely maximizing their long-term self-interest by incorporating expected future reciprocity. This reduces altruism to enlightened self-interest.[5]

Alternatively, the problem can be resolved by redefining the self that is constituted by self-interest. Sociobiologists have argued that individual fitness refers to genetic fitness and, therefore, that altruism toward kin increases genetic fitness.[6]

---

uals distinguish between opportunity costs (forgone gains) and real costs and are more willing to part with some potential gains than to incur real losses. For another formal characterization of different kinds of benevolence, see Stefan Valavanis, "The Resolution of Conflict When Utilities Interact," *Journal of Conflict Resolution* 2 (June 1958): 156–69.

4. Adam Smith, the exponent of laissez faire and the invisible hand, did understand the role of sympathy rather than self-interest: "How selfish soever man may be supposed, there are evidently some principles in his nature, which interest him in the fortunes of others, and render their happiness necessary to him, though he derives nothing from it, except the pleasure of seeing it." Quoted in Wilson, "Sympathy and Self-Interest," p. 73.

5. This is the strategy taken by economists. See, for examples, Peter Hammond, "Charity: Altruism or Cooperative Egoism?" in *Altruism, Morality, and Economic Theory*, ed. Edmund S. Phelps (New York: Russell Sage Foundation, 1975), pp. 115–31, and Gary S. Becker, "Altruism, Egoism, and Genetic Fitness: Economics and Sociobiology," *Journal of Economic Literature* 14 (September 1976): 817–26. Biologists have also taken the route of reducing altruism to the egoistic expectation of future reciprocity. See Robert L. Trivers, "The Evolution of Reciprocal Altruism," *Quarterly Review of Biology* 46 (1971): 35–57.

6. See the pioneering article by William D. Hamilton, "The Genetical Evolution of Social Behaviour," *Journal of Theoretical Biology* 7 (1964): 1–32. Also see Edmund O. Wilson, "The Genetic Evolution of Altruism," in *Altruism, Sympathy, and Helping: Psychological and Sociological Principles*, ed. Lauren Wispe (New York: Academic Press, 1978), pp. 11–37, and Mark Ridley and Richard Dawkins, "The Natural Selection of Altruism," in *Altruism and Helping Behavior: Social, Personality, and Developmental Perspectives*, ed. J. Phillipe Rushton and Richard M. Sorrentino (Hillsdale, N.J.: Lawrence Erlbaum Associates, Publishers, 1981), pp. 19–39.

7

Conclusion: Structure, Circumstance, and Choice in International Relations

International cooperation and conflict are inextricably joined. Both are omnipresent in world politics, as they are in many other forms of social and biological relations. Both reflect purposive calculated behavior in an interdependent world, and both emerge from the interaction of an array of situational factors. The arguments developed in this book start from an understanding that the existence of separate sovereign states concerned with their survival underlies the gamut of relationships from amity to enmity.[1]

Yet scholars typically consider either cooperation or conflict alone. Those who focus on conflict all too often ignore the cooperative elements of international relations. Moreover, competitive and conflictual relations can underlie concerted, cooperative ones. A world of only defensive weapons—and, therefore, without military rivalries—would also be one without allies and joint military action. International cooperation is embedded within a structure of competition, rivalry, and insecurity. In an anarchic and conflictual world, states develop and nurture cooperative relationships.[2] Tacit bargaining can even occur between enemies in the midst of war.

A focus on cooperation alone is equally problematic. Some econo-

---

1. See Arnold Wolfers, "Amity and Enmity among Nations," in *Discord and Collaboration: Essays on International Politics* (Baltimore: Johns Hopkins Press, 1962).

2. Analogously, markets and exchange presume scarcity; biological competition undergirds kin altruism; and cooperative helping behavior among people is exhibited during disasters.

173

mists recognize that their discipline focuses on mutually beneficial exchange and that the efficiency of the market masks a great deal of the
coercion, power, and conflict that attend market exchanges.[3] Similarly,
a focus on international institutions and norms underemphasizes the
role of conflict in world politics. Even cooperative interactions have
conflictual and competitive elements.

Outcomes of international cooperation and conflict emerge as a result of states' strategic choices, which include both cooperation and
conflict as strategies. Nations are neither inveterate cooperators nor
defectors. Both options constitute parts of states' repertoires of behavior, and countries use both to ensure survival and fulfill national interests. Thus both stem from the same source.

All states are capable of allying with others as well as waging war on
them. States can and do shift between strategies of alignment and nonalignment. Even those states with long histories of nonalignment have
the capacity for cooperation. The United States avoided entangling alliances for more than a hundred years, but it did enter two wartime
alliances in the first half of this century. Then, in 1949, for the first time
in its history, the United States entered a peacetime alliance, NATO.
Great Britain also eschewed alliances during the nineteenth century
but departed from its policy of splendid isolation early in the twentieth.

Sometimes choices are clear-cut. Purposive calculation in some contexts generates a single strategic choice. At other times, however, states
are faced with a dilemma: they need to decide which strategy to follow,
but they find a compelling logic and rationale for each. The criteria
that generate incommensurate strategies can be explicated, and the
situations in which they arise can be delineated.[4]

The existence of dilemmas of choice means that there are different
ways to assess self-interest. Rationality and self-interest are not necessarily unambiguous guides to, or explanations of, action. National

3. The work of Jack Hirshleifer provides an excellent example. See his collected essays,
Economic Behaviour in Adversity (Chicago: University of Chicago Press, 1987).
4. Philosophers and social scientists have devoted a good deal of scholarship to resolving dilemmas of choice. But dilemmas can only be explicated, not resolved. To demonstrate that contingent cooperation emerges as individualistically rational in an iterated
prisoners' dilemma with sufficient weight attached to future payoffs is only to demonstrate the contextual bounds within which it is a dilemma. The dilemma is not resolved;
it is shown only to exist in particular circumstances. Other solutions entail making certain
decision rules inadmissible (a technique of moral philosophers) or making certain options
impossible (a technique of political theorists). Coercion represents a way of dealing with
the dilemma but does not solve it. Drugs that deny people the ability to think about
anything but immediate self-gratification resolve intertemporal dilemmas, for example,
but do so by making impossible one way of calculating self-interest.

interests are not always self-regarding. A state's preferences derive from assessments of the payoffs of other nations as well. Self-interest can be assessed individualistically, relatively, or conjointly. Self-interest can be maximized in the short term or in the long term. Sometimes these criteria—the bases for calculating self-interest—are incompatible.

Dilemmas of choice between cooperation and conflict mean that rational self-interested behavior can be self-defeating.[5] In addition, individualistically rational behavior can be collectively irrational.[6] Short-run rationality can result in long-run disaster. A concern with relative or joint gains can entail forgone absolute payoffs, whereas a focus on absolute payoffs can destroy one's relative position and one's relationships with friends and allies. What is in an actor's self-interest by one way of reckoning runs counter to its self-interest by a different measure.

Realism, Liberalism, and International Cooperation

International cooperation and conflict result from choice and assessment; they are the products of payoffs, perceptions, and bases of calculation. All these factors are crucial; all can have a role in determining international relations. Yet not all are addressed in the corpus of realist and liberal thought.

Liberals and realists agree that states cooperate because it is in their interest to do so. To liberals, such behavior is the norm. To realists, on the other hand, international cooperation is rare and transitory. Both these contradictory conclusions imply specific assumptions about the constellation of payoffs that underlie the behavior of nations. Since cooperation emerges when it is mutually advantageous, liberals must presume that states have common interests. As a result, self-interested interaction leads to international cooperation. In contrast, realists see a world of conflicts of interest. But if international relations involve both common interests and conflicting ones, both liberals and realists are both right and wrong.

---

5. Derek Parfit characterizes the prisoners' dilemma as exemplifying how self-interest can be indirectly self-defeating. See Derek Parfit, *Reasons and Persons* (Oxford: Oxford University Press, 1986). For me, the issue is less one of directness and indirectness than of the temporal and cross-sectional basis by which utilities are calculated and assessed.

6. Issues of collective irrationality arise only when all actors confront a dilemma. Some of the dilemmas discussed are individual ones that pose a quandary for one actor. Some situations pose dilemmas for all actors. For delineations of such social dilemmas, see Robyn M. Dawes, "Social Dilemmas," *Annual Review of Psychology* 31 (1980): 169–93, and Wim B. G. Liebrand, "A Classification of Social Dilemma Games," *Simulation and Games* 14 (June 1983): 123–38.

The two perspectives differ in their treatment of perception. Realism presumes no misperception; it assumes an unambiguous environmental imperative. States in the anarchic setting of international politics cannot misconstrue the nature of their predicament and the threats that emanate from others. This failure to deal with the implications of misperception is a weakness in realist theory. Unless a state's choices are never contingent on the choices of others, misperception can matter. But the situations discussed by realists, such as chicken, are ones in which a state's behavior is contingent. By contrast, liberals emphasize the role of perception in their arguments but do so one-sidedly. They argue that absent misperception, cooperation would necessarily emerge. But as already discussed, this presumes that cooperation is the benchmark norm from which deviations are generated by misperception. Thus we are back at the problem that liberalism presumes the absence of situations of pure conflict.

The two perspectives also differ in the bases of calculations that underlie actors' choices. Both perspectives stipulate the role of self-interest but do not adequately address the ways in which interests are assessed. Liberalism presumes egoistic and individualistic self-interest. In contrast, the realist emphasis on power would suppose a competitive and relativistic assessment of payoffs, something that remains ambiguous in much realist thought. Moreover, as discussed in Chapter 4, the realist emphasis on the core objective of survival can also imply different bases of calculation.

For the bases of calculation vary as a function of circumstance and relationships. States may not be able simultaneously to make relativistic assessments vis-à-vis all others in the international system. Moreover, the fact that in some situations decision criteria generate different strategies implies that states must make metachoices—they must choose how to go about choosing. All this means that liberalism and realism may both be correct, but about different relationships and circumstances.

Levels of Analysis and the Explanation of International Cooperation and Conflict

The conclusions about the centrality of strategic interaction entail analytic implications for the study of international relations. In addition to constituting a set of substantive conclusions about the underpinnings of international cooperation and conflict, they imply that the standard approaches to explanation in international politics can be incomplete, and they suggest a basis for integrating alternative perspectives.

A key analytic conclusion, and one that seems quite obvious, is that strategic interaction is important.[7] The arguments developed in this book all demonstrate that cooperation and conflict are products of circumstance and choice. Decision is a product of a state's options, payoffs, and criteria for calculation, conjoined with the situations confronting other nations. Those situations matter either because a nation's choice is contingent or because others' payoffs are incorporated into the calculus. Choice, then, involves interaction and the forces of circumstance.

Such a view differs from the standard wisdom for explaining international politics. In this section I categorize scholarship in the field and argue that various approaches are incomplete because they do not focus on interaction.

Scholars often categorize international politics by level of analysis, referring to first-image arguments (those operating at the individual level of analysis), second-image arguments (those dealing with the national level of analysis), and to third-image arguments (structural explanations at the systemic level of analysis).[8]

Individual Level of Explanation

One hallowed approach to international politics focuses on individuals. Since nation-states are governed by people, it treats international politics as a function of human behavior and a product of human choice. It assumes that the same factors that determine individual decisions

7. This book uses a strategic-interaction approach and so presumes that strategic interaction matters. On the other hand, the concrete conclusions about the bases of cooperation demonstrate that strategic interaction does indeed matter in specific circumstances.

8. The three-images classification comes from Kenneth N. Waltz, *Man, the State, and War: A Theoretical Analysis* (New York: Columbia University Press, 1959). Waltz's images were put in levels-of-analysis terms in a review essay by J. David Singer, "International Conflict: Three Levels of Analysis," *World Politics* 12 (April 1960): 453–61. Singer's views are fully expounded in J. David Singer, "The Level-of-Analysis Problem in International Relations," *World Politics* 14 (October 1961): 77–92. More recently, Robert Jervis has proposed using four levels of analysis. See Robert Jervis, *Perception and Misperception in International Politics* (Princeton, N.J.: Princeton University Press, 1976), chap. 1. Elements of my view of the levels of analysis are to be found in Arthur A. Stein, "Restraints, Imperatives, and the Analysis of Soviet Foreign Policy," paper presented at a conference on Domestic Sources of Soviet Foreign and Defense Policy, University of California, Los Angeles, October 1985; "Structure, Purpose, Process, and the Analysis of Foreign Policy: The Growth of Soviet Power and the Role of Ideology," unpublished manuscript, 1986; and "Constraints and Determinants of Decision Making: Structure, Purpose, and Process in the Analysis of Foreign Policy," manuscript in progress.

explain national ones as well; in other words, people wage war and make peace for the same kinds of reasons they make decisions about anything else.

The roots of interstate rivalry may be understood to lie, for example, in the aggressive nature of particular human beings. Personality theories, one form of individual-level explanation, link international conflict to the belligerent traits of specific political leaders.[9] They reduce the focus of explanation from nation to individual and from a relationship to the actions of a single person. At their crudest, such explanations ignore the actions of other states altogether; they neglect the international environment completely. They hold, for example, that individuals with certain upbringings will act aggressively, and if they happen to become leaders of states, national aggression may result. Hitler, seen as a madman, is thus understood to have gone on an aggressive spree that caused World War II.[10]

These studies reduce the study of international politics to psychohistory and view the foreign policy of a state as the direct expression of a leader's personality and as unaffected by the actions of others. Such analyses do not even clarify whether these leaders themselves analyze other countries except as objects of aggression (the international relations equivalent of individuals who look upon other people as sex objects). In this view the personal experiences and upbringings of political leaders affect politics on any level, international, national, or local. Personality approaches deal with neither nations nor relations, therefore.[11]

9. Saul Friedlander and Raymond Cohen, "The Personality Correlates of Belligerence in International Conflict: A Comparative Analysis of Historical Case Studies," *Comparative Politics* 7 (January 1975): 155–86.

10. The view of Hitler's behavior as aberrant is so common that outrage greeted A. J. P. Taylor's revisionist thesis that Hitler was just another statesman using the opportunities provided by other nations to advance his nation's interests. See Taylor, *The Origins of the Second World War*, 2d ed. (Greenwich, Conn.: Fawcett Publications, 1961); William Roger Louis, ed., *The Origins of the Second World War: A. J. P. Taylor and His Critics* (New York: John Wiley, 1972); Gordon Martel, ed., *The Origins of the Second World War Reconsidered: The A. J. P. Taylor Debate after Twenty-five Years* (London: Allen and Unwin, 1986). John Mueller argues not only that Hitler is a necessary and sufficient explanation for World War II, but that World War I was the high water mark for war among the major powers and that attitudes in these societies meant that they would, from then on, have abjured war with one another, if not for the aberration of Hitler. See Mueller, *Retreat from Doomsday: The Obsolescence of Major War* (New York: Basic Books, 1989).

11. Such trait explanations do not adequately explain behavioral changes over time and across space. Individuals predisposed to act aggressively should do so constantly and should lash out indiscriminately. To explain why aggression is observed only sporadically, or why it is directed at only some targets, requires the incorporation of additional, typically situational, determinants. It is possible to construct richer personality

Such trait explanations deny the importance of interaction and strategy. They analyze why individuals act, but they take no account of interaction. They view behavior as a function of each actor's feelings and predispositions, not as one's response to the behavior of another. Almost universally, they treat cooperative or conflictual behavior as the outgrowth of an actor's intrinsic characteristics.

National Level of Analysis

Other analysts treat international politics as a function not of individual but of national characteristics. A perspective with a long pedigree, this view has been variously dubbed the second-image argument or national level of analysis. It sees states as entities whose behavior, like that of individuals, can be explained with reference to their individual natures. Large states act differently from small states, rich nations act differently from poor ones, and democratic or libertarian states act differently from authoritarian or totalitarian ones.

This particular form of national-level explanation shares the problems inherent in individual-level analyses. It denies the importance of interaction and strategy. Almost universally, it treats cooperative or conflictual behavior not as one's response to the behavior of another, but as the outgrowth of an actor's intrinsic characteristics.[12]

Another set of national-level arguments, organizational process models of foreign-policy decision making, also ignores international interaction.[13] These explanations, which treat international politics as a direct outgrowth of domestic politics, view foreign policy as the result of a domestic tug of war; the only interactions they consider are those between vying domestic agencies, whether armed services or civilian bureaucracies. The protectionists in the Commerce Department battle the free traders in the Office of the Special Trade Representative, for example, and State Department Arabists vie with pro-Israelis on the White House staff.

---

explanations, ones that do have some contingency in their explanations. See, for example, Alexander L. George and Juliette L. George, *Woodrow Wilson and Colonel House: A Personality Study* (New York: J. Day, 1956); Thomas M. Mongar, "Personality and Decision-Making: John F. Kennedy in Four Crisis Decisions," *Canadian Journal of Political Science* 2 (June 1969): 200–225; and Lloyd Etheridge, "Personality Effects on American Foreign Policy, 1898–1968: A Test of Interpersonal Generalization Theory," *American Political Science Review* 72 (June 1978): 434–51.

12. Fixed national characteristics also do not adequately explain behavioral changes over time and across space. States predisposed to act aggressively should do so constantly and should lash out indiscriminately. Here, too, explaining why aggression is observed only sporadically, or why it is directed at only some targets, requires the incorporation of additional, typically situational, determinants.

13. These processes are often referred to as bureaucratic politics.

By failing to incorporate any sense of international interaction, this approach, like the other forms of explanation discussed thus far, fails to provide any sense that decisions and calculations come in response to, or in a context of, an ongoing relationship that constrains and shapes interests, choices, and behavior. Instead, agencies are seen as automata that employ standard operating procedures and, as a result, respond similarly to comparable situations. Indeed, process models often seem to treat people and organizations as servomechanisms. Like thermostats, which monitor temperature and have only a small repertoire of responses, people and bureaucracies monitor a small aspect of an environment and can respond with only a limited, pre-programmed repertoire. In such models the people and organizations within a nation interact with one another but are not seen as having any relationships with or any ability to calculate strategy toward other nations.

Rational-Choice Arguments

Finally, at both levels of analysis there are rational-actor models. At the individual level, foreign policy can be explained not only by reference to personal traits but as a function of human rationality and personal choice. At the national level, policy can be understood as an expression of a state's corporate interests that emerges from purposive behavior intended to advance specific objectives.[14]

Unlike trait explanations, rational-choice arguments, whether applied to the individual or the nation, do incorporate situational elements in the explanation of behavior. What a person or a country will do is understood to vary from one time to another and from one case to another as available options shift or as more information about the likely consequences of different actions becomes available. Behavior is not seen as driven by an undifferentiated notion of trait.

14. For synopses of the use of rational-actor models in international relations, see Graham T. Allison, *Essence of Decision: Explaining the Cuban Missile Crisis* (Boston: Little, Brown, 1971) and John D. Steinbruner, *The Cybernetic Theory of Decision: New Dimensions of Political Analysis* (Princeton, N.J.: Princeton University Press, 1974). Rational-choice explanations for both individuals and corporate entities have a long history in the social sciences. Such rational-choice explanations of international politics are much the same as economic explanations of individual and firm behavior. In both, actors—be they individuals, firms, or states—are assumed to have fixed and hierarchically arranged objectives, to have full information about options and their possible consequences, and to maximize expected utility. What distinguishes international relations theorists from economists is not only their focus on states rather than firms, but also the utility they assume the actors in question are maximizing. Economists talk of maximizing profits, whereas political scientists talk of maximizing power or security.

Yet many rationalist explanations also fail to consider interaction. They depict individuals and states as having preferences, confronting options, and making optimal decisions. Such explanations posit interests and bases of calculation. Assuming that states have full information and want only to maximize security, for example, allows the analyst to move from context to outcome without really focusing on the intervening steps of calculation, interaction, and decision. Varying conditions explain variations in behavior.[15] By treating the international environment as a market, these explanations draw from the economic model of a perfectly competitive world in which firms base their decisions on assessments of the likely demand for goods and without regard for the expected behavior of other companies. Firms neither cooperate nor compete with one another directly.[16] Rather, they vie anonymously and indirectly with all others in the market, although they neither affect nor are affected by others' actions.[17]

The appropriate analogue for international politics is not a perfectly competitive market, however, but oligopoly or imperfect competition— a domain of small numbers with a few dominant firms in which each actor's decisions are contingent on the choices of others. Hardly oblivious to one another, actors compete directly and collude with others. They make decisions in full awareness of others. In short, an analysis of their behavior must take account of their strategic interaction.

Systemic/Structural Explanations

Systemic explanations of state behavior do recognize the centrality of interaction and context. Indeed, interaction provides the defining characteristic of structural or systemic explanations. Nonetheless, because interaction in these models is fully context-dependent, it plays no role in explaining outcome. Again, extant traits provide the full explanation. Here, however, the critical characteristics are extrinsic, belonging not to the actors but to the environment in which they operate. Hence a bipolar world is thought to have certain consequences, a multipolar one others. The interactions of states follow directly from, indeed are dictated by, the international distribution of power.

15. The intervening steps of calculation and choice are posited, not studied, and they are invariant.

16. Economists, for example, posit that firms are profit maximizers and have full information. They can then explain and predict the behavior of firms solely by assessing antecedent conditions.

17. If the international environment were like a marketplace, individuals and states would confront generalized contexts (market conditions) but not other leaders or other states.

To link global conditions and outcomes, structural and systemic explanations explicitly posit the nature of actor interests.[18] Balance-of-power theorists, for example, typically assess the nature of state interests either by positing that these are quite general in character and neither historically nor geographically specific or by deducing them from the distribution of power. Examples of the former arguments are those holding that all states seek minimally to ensure survival or maximize power. An example of the latter is that when the global distribution of power is skewed—when one great power dominates others—this hegemon necessarily finds free trade to be in its interest.[19]

Indeed, those who use systemic models posit state interests to deduce the consequences of structural configurations and systemic changes on international politics. Hence, assuming that states maximize some utility, it becomes possible to deduce how they would respond to changes in the global distribution of power. Having assumed the intervening step of state interests and calculations, scholars can move blithely from the distribution of power to assessing a particular outcome. Positing state interests makes it possible to neglect the intervening stage in which states interact with one another.

Such structural theories are sometimes erroneously characterized as macro theories that provide no micro explanations. Kenneth Waltz characterizes structural explanations as theories of international politics rather than theories of foreign policy.[20] In other words, such theories presume to explain general patterns and not the specific behavior of individual states.

Yet structural theories must have micro foundations. In fact, structuralists posit that states rationally pursue their national interests, whether defined in terms of security, power, wealth, or prestige. And because the behavior of the small set of great powers shapes the international system, structuralists argue, it is the specific behavior of these states that structural theories must perforce explain. Hence, if systems theorists argue that balances of power emerge, they are necessarily suggesting, as well, how great powers will act in specific situations so that balances will emerge.[21]

18. As such, they are intimately linked to rational-choice arguments described later in this chapter.
19. Stephen D. Krasner, "State Power and the Structure of International Trade," *World Politics* 28 (April 1976): 317–47.
20. Waltz's choice of title, *Theory of International Politics*, is quite telling.
21. Waltz specifically berates other theorists for being reductionists and not true structuralists. Yet, as I argue, structuralist theories for small sets of nations necessarily entail reduction. This is especially true for bipolarity, which focuses on the behavior of two

A central argument of this book is that the nature of states' calculations must be more precisely stipulated. Because nations confront situations in which there are compelling logics for opposed courses of
action, it may be impossible to ascertain their behavior without knowing their decision criteria and that they perceive accurately. Since structural theories do not treat either the bases of calculation or perception
as variable, they must be supplemented—the specifics of strategic interaction between states must be assessed.

Because interaction does matter, systemic explanations are incomplete. They assume, for example, that given a particular distribution of
power, a great power will pursue a particular path, a middle power a
different one, and a lesser power still another. Yet strength does not
dictate strategy. Sometimes, for example, weaker nations appease opponents, and sometimes they balance them; they can attempt to ingra-
tiate themselves with opponents or confront them.[22] Obviously, mere
position and relative power do not tell us about particular patterns of
alignment. There remains some strategic indeterminacy after structural
factors have been analyzed.[23]

The international system structures, but does not determine, choice
and outcome. In this book I have demonstrated the interlocking roles
of payoffs, perceptions, and the circumstances that underlie different
decision criteria in determining the choices that generate international
cooperation and conflict. International-systems explanations can be
complete only if they determine all these factors or if the factors they
do not determine also do not matter. But systemic factors cannot determine what misperceptions actors may have and what decision criteria they will adopt. Thus systemic factors are complete only if they
determine the payoffs which, in turn, solely determine choice and
outcome.

The theory of hegemonic stability provides an excellent example of
a structural argument that, to its peril, ignores strategic interaction.

great powers. Waltz criticizes the reductionism of others, argues that he has no preten-
sions about presenting a theory of foreign policy, and yet proceeds to describe how
specific states must react in a bipolar world.

Although Waltz's systemic theory does include a theory of foreign policy, it is incomplete. It explains how an equilibrium is restored through a balance of power once the
international situation is disturbed, predicting, for example, how great powers react when
one of them embarks on a course of global conquest. The theory cannot explain why
some states choose such a course, however.

22. Brian Healy and Arthur Stein, "The Balance of Power in International History:
Theory and Reality," *Journal of Conflict Resolution* 17 (March 1973): 33–62.

23. Interestingly enough, we see no contradiction in saying that states surrounding
the Soviet Union look to the United States as a counterweight and protector, whereas
Canada and Mexico do not similarly turn to the Soviet Union.

Used to explain the existence of international regimes in various issue areas, this theory is a variant of the classical balance-of-power theory applied to the arena of international political economy.[24] Hegemony refers to a state's position relative to others; a hegemon is more powerful than other states and stands alone without equal at the top of the international hierarchy of power. The theory posits that the existence of a hegemonic power leads to the emergence and maintenance of stable regimes. The model is systemic in that it begins with a particular distribution of power across the set of nation-states in the international system. The theory derives from a deduction that free trade provides the maximum absolute payoff for a hegemonic power. But it presumes the calculations that the hegemon will make and has no component of strategic interaction. It provides no basis for the unstated but essential supposition that such states will abjure nonindividualistic calculations. Further, it provides no sense of the process by which the existence of a hegemonic distribution of power leads to the particular set of rules that define the international regimes in question.[25]

States interact and negotiate, and international outcomes result from these processes. Explaining international outcomes requires more than the mere specification of states' interests given a particular distribution of power. It also requires a knowledge of the nature of states' interactions given particular constellations of interests and power. This book maintains that there are intervening steps of calculation, bargaining, and strategy that come between the structure of the international system and the choices of states and that result in interactions and outcomes.

Strategic Interaction as a Level of Analysis

The independent causal import of the circumstances underlying strategic interaction suggest that it should be understood as constituting

24. Arthur A. Stein, "The Hegemon's Dilemma: Great Britain, the United States, and the International Economic Order," *International Organization* 38 (Spring 1984): 355–86.

25. The theory has been criticized for this very reason. It assumes either that a hegemon induces other states to join a free trade regime or that it forces them to do so. The induced argument is problematic because the hegemon must negotiate a trade regime with states that may not find it to be in their interests. The coercive argument is problematic in that the hegemon can only coerce minor, not major, trading powers. For my views, see Stein, "The Hegemon's Dilemma." For an excellent review, see Duncan Snidal, "The Limits of Hegemonic Stability Theory," *International Organization* 39 (Autumn 1985): 579–614. For another argument that includes some aspects of interaction, see David Lake, "International Economic Structures and American Foreign Economic Policy, 1887–1934," *World Politics* 35 (July 1983): 517–43. For a theory that includes not only a logic of interaction but an explanation for cycles of hegemony, see Mark R. Brawley, "Challenging Hegemony: How Cycles of Hegemony, Hegemonic Decline, Major War, and Hegemonic Transitions Are Linked," Ph.D. dissertation, University of California, Los Angeles, 1989.

another level of analysis. International outcomes are determined by the choices of states which are, in turn, affected by the circumstances of strategic interaction. This component of interaction is neither captured nor subsumed by the other levels of analysis. Moreover, a focus on strategic interaction makes it possible to integrate the other levels of analysis and to delineate both the roles they play and the situations in which they matter.

Systemic forces structure but do not necessarily determine national choices. They delineate the context in which states interact. They may shape the options and payoffs that states confront. Yet states transform those payoffs into utilities using different decision criteria. The international system sets a framework for interaction but only rarely dictates specific strategies in specific circumstances.

It is in situations of strategic indeterminacy that the other levels of analysis become important.[26] Individuals or nations with certain characteristics may be more likely to pursue competitive rather than individualistic gains. As discussed above, there are circumstances in which orientation is critical in determining choice. And as discussed further below, cases of systemic indeterminacy, in which decision criteria point toward different strategies, also underlie domestic political debates. Internal political factors can be important in choosing foreign policy strategies when there are competing logics for assessing the national interest. Further, there are situations in which misperception matters, in which a knowledge of the cognitive or organizational underpinnings of inappropriate information processing is a central component of explanation.

A strategic-interaction approach can bridge the gap between systemic and decision-making approaches to the study of international relations. It makes it possible to delineate the situations in which decision making may matter and the ways in which it matters (in the choice of decision criteria or in particular forms of misperception).

---

26. Some scholars mistakenly believe that they control for structure by comparing national responses to the same event. They believe that they make the case for the causal role of domestic politics, for example, by comparing national responses to the oil crisis of the 1970s. The argument made in this book is that structure generates constraints and opportunities, delineating options and payoffs. Unless these are the same for a set of nations, systemic forces have not been controlled for. The same oil shock has different implications for oil exporters than for oil importers, as it does for oil importers with varying domestic capacities, and so on. The same international event can generate different opportunities and constraints for different actors. Domestic politics clearly plays a determining role when structural factors generate the same options and payoffs for actors who respond differently.

Dilemmas of Immediate Egoism

Nations interested in their own immediate payoffs (egoists) can confront dilemmas of strategic choice. Although structure is a key determinant of the possibilities and payoffs for states, these do not, in
turn, necessarily determine outcome in and of themselves.[27] Assuming that foreign policy is purposive behavior explicable by reference
to objectives and calculations, any such explanation presumes that
payoffs are accurately perceived and that the comparison is straight-
forward. How states perceive, compare, and translate those payoffs
into utilities plays a critical role. Yet states, like people, may choose
one strategy over another because they miscalculate, misperceive, or
because of the way in which they transform and translate payoffs
into utilities.

States may depart from the imperatives of the payoff environment
when they misperceive. Yet misperception matters only when states'
choices are contingent on the choices of others or when dominant
choices generate suboptimalities or leave aggrieved actors attempting to improve their expected outcomes. More important, misperce-
tion can foster coordination and cooperation as well as generate
conflict.[28]

States confront situations in which autonomous decision making can
lead to suboptimal and undesirable results which generate dilemmas
of common interests and common aversions. Actors who foresee such
possibilities have an incentive to eschew independent decision making.
Hence they create regimes to ensure outcomes that are mutually desirable or to avoid outcomes that are mutually undesirable.

Two critically important points emerge. First, individualistic, self-
interested, autonomous behavior can lead to cooperation as well as
conflict. Second, some situations generate dilemmas in which there
exists a logic for either of two available courses of action.

27. For the relationship between structure and other factors as both constraints and
determinants of policy, see Arthur A. Stein, "Restraints, Imperatives, and the Analysis of
Soviet Foreign Policy"; "Structure, Purpose, Process, and the Analysis of Foreign Policy";
and "Constraints and Determinants of Decision Making."

28. Accurate perception and full information of payoffs do not always generate unambiguous imperatives, however. An accurate and complete knowledge of the payoff
environment does not necessarily dictate the choice between cooperative and conflictual
strategies. Dilemmas remain. Since states can be aggrieved even with outcomes they
expect, they can opt for strategies, when available, to improve on those expected outcomes. States, in short, embark on strategies seemingly not in their interests in order to
improve their payoffs. Elsewhere, I have discussed strategies of linkage in such terms;
see Arthur A. Stein, "The Politics of Linkage," *World Politics* 33 (October 1980): 62–81.

Dilemmas of Prospective Egoism: Future versus Present

States also confront trade-offs between their current payoffs and their future ones. Sometimes they choose to forgo current rewards for long-term ones; sometimes they do not.[29] This is one source of a potential dilemma in which a short-term logic suggests the superiority of either cooperation or conflict and a long-term logic favors the other. For focusing on the long term does not guarantee cooperation even in the prisoners' dilemma.

Intertemporal trade-offs can loom large for nations dealing with prospective payoffs. Here the choice between cooperation and conflict can in certain situations be determined by nations' current positions, their attitudes toward risk, and the discount rates they attach to future payoffs. States whose survival is ensured can afford to think of the long term and accept risk. Those less secure, more likely to think in terms of immediate payoffs, will avoid risk. They may choose not to maximize expected payoffs, but to concentrate on attaining sure things. They may approach their values lexicographically and choose to maximize their chances of survival without regard to other national values. Assuming that survival is ensured, states may take greater risks in the domain of losses than in the domain of gains.

To summarize: a state's degree of security and its potential payoffs combine to define its attitude toward risk. That perspective in turn determines how a state calculates in choosing between a cooperative and a conflictual strategy. In this way temporal horizons can be determinative in international politics.

Dilemmas of Rivalry and Entanglement

Dilemmas of choice between strategies of cooperation and conflict also arise because self-interest is not synonymous with egoism. A state's payoffs do not necessarily translate directly into utilities and so do not dictate a particular strategic choice. Self-interested behavior need not be solely self-regarding; it can be other-regarding as well. Self-interested calculated choice is consistent not only with individualistic but also with competitive or conjoint decision criteria. Because choice entails an intervening step in which nations assess payoffs and transform them

29. For a discussion of debates about such trade-offs in economic policy, see Amy E. Davis, *The Kennedy Presidency and the Politics of Prosperity* (Cambridge: Cambridge University Press, forthcoming).

into utilities, the criteria by which decisions are made matter. How a nation calculates its self-interest, whether it adopts an individualistic, competitive, or conjoint orientation, affects its choice of strategy.

Moreover, the character of the relationship between states also underlies the nature of their strategic interaction. A prisoners' dilemma between allies differs from one between enemies. States can interact with others as individualists concerned only with their own payoffs, as rivals concerned with relative gains, or as allies concerned with joint gains. States may or may not have preferences regarding others' payoffs.

States may also calculate utilities on the basis of relative payoffs. There does exist a self-interested logic for the use of a competitive decision criterion, for maximizing relative outcomes rather than absolute ones. States do not compete with all others, only with certain ones. The combination of geopolitics, technology, and power determines rivalries.[30] Rivals calculate self-interest competitively, and their conflicts of interest make cooperation impossible.[31]

When a state must choose between an individualistic and a competitive strategy, it confronts a dilemma of rivalry. It must decide whether to eschew absolute returns for relative position. If it focuses on relative returns, it will exacerbate conflicts of interest that already exist. Those conflicts of interest are inherently constant-sum when they involve goods whose quantities are fixed. Territorial disputes provide one striking example. Other phenomena, such as weapons deployment, involve no naturally fixed quantity of goods but become constant-sum in character because states approach them using competitive decision criteria.

Arms races, because they involve such relativistic competition, are not typically amenable to negotiated solutions. Two states locked in such a race share, at most, a common interest in restraining third countries or in channeling their competition. As the historical record demonstrates, searches for negotiated solutions to arms competitions will be ill-fated unless one of the parties accepts relative inferiority. States use negotiations as extensions of their arms-racing behavior, as a means of constraining their opponents. Typically such negotiations fail, channel the arms race toward a particular category of weapons, or merely affirm that one side has accepted relative inferiority.

30. This is discussed later in the chapter.
31. It is also possible to conceive of situations in which actors attach no weight to their own payoffs and focus solely on minimizing the payoffs to others. States that embark on holy wars and crusades are sometimes unconcerned with their own gains but focus solely on transforming others. Ideological contests in international relations sometimes develop this flavor. In such cases a state's utility derives solely from converting the infidel.

Yet self-interest can also lead to the use of concerted decision criteria. International relationships run the gamut from intense rivalries to close alliances. Some alliances, mere marriages of convenience, codify temporarily aligned interests and in no way constrain the member nations' pursuit of egoistic self-interest. In addition to rivalries and these generally short-term coalitions, however, are alliances in which states pursue joint interests. They enter these arrangements as autonomous entities in order to ensure that others accord their concerns some weight, and they promise to take those others' needs into account when making their own decisions. Such alliances are not federations that create unified and integrated decision-making arrangements. Their members may have conflicting as well as shared concerns.

In such alliances, dilemmas of choice arise when individualistic self-interest conflicts with the use of a decision criterion emphasizing joint concerns. This is the alliance dilemma, the need to decide whether to pursue self-interest individualistically or jointly, whether to restrain egoistic concerns in favor of the needs of an ally. The use of conjoint decision criteria and an emphasis on joint interests is no guarantee of cooperation. It only ensures that such relationships will not collapse when the slightest conflict of interest arises.

Self-interest is certainly one basis for the use of both competitive and conjoint decision criteria. The former does not represent spite, and the latter does not equal altruistic self-denial. In both cases states attach weights to their own payoffs, but their utilities, and thus their decisions, are not based solely on their own payoffs.

Further, one can argue that states merely downplay their immediate self-interest and attach greater weight to prospective payoffs—that they look to the future and incorporate potential payoffs into their current calculus. If this is so, the basis for competitive rivalry lies in states' desires for the payoffs that will be realized when others have been driven from the game. Alternatively, the rationale for immediate sacrifice derives from their expectations of the gains that will come from maintaining an alliance. Since states do not know what those future payoffs will actually be, however, they base their decisions on their appraisals of current payoffs. In other words, it is important to distinguish between the expectations of future interactions and their payoffs. Even when states anticipate a future relationship, they do not know what the payoffs associated with it will be. A commitment to sustain an alliance does not guarantee that it will not founder.

A state chooses to sustain a relationship with unknown future payoffs based on a set of concrete payoffs and demands today. The United

States did not know at the time of the Nassau summit in 1962 what the payoffs of maintaining the alliance with Britain would be. It did not even know whether a decision unfavorable to British interests would jeopardize the coalition. It only knew what Britain needed and what filling that need would cost the United States. Put differently, little in life resembles a prisoners' dilemma game played ad infinitum with exactly the same payoffs for each game. All a state knows at any given time is what will maximize its relative gains immediately and what the cost will be to its allies if it pursues egoistic payoffs.

This argument does not deny the importance of self-interest. It is not about spite, envy, kinship, or altruism. Nor does it deny the role prospective payoffs can play in making decisions about immediate actions. Rather, because future returns are typically unknown, states choose what to do on the basis of current relative or joint payoffs. Different criteria for choice are thus compatible with self-interest. And self-interest can be the basis for competition and conjunction.

Dilemmas, Interdependence, and Extended Preference

Dilemmas of strategic choice entail the existence of compelling logics to cooperate and to conflict. They arise when interaction involves both common and conflicting interests. The existence of potential outcomes in which one state benefits when another does not generates a conflict of interest and an imperative for conflict. The possibility that both may benefit, on the other hand, creates a common interest and a logic to cooperate. The dilemma of which strategy to choose arises only when states' decisions affect not just their own payoffs but those of others as well. In short, dilemmas of strategic choice are rooted in interdependence and imply that states have preferences as to the choices of others—that they are not indifferent to what others will do.

The dilemmas of self-interest delineated in this book have different bases. Some dilemmas are rooted in the absence of a dominant strategy. In such situations of contingency, it is not surprising that there are logics associated with different strategies. But even actors with dominant strategies confront dilemmas. The prisoners' dilemma, for example, is rooted in the conflict between dominance and optimality. In still other cases dilemmas are rooted in temporal conflicts between short- and long-term maximization.

Dilemmas of strategic choice are also rooted in utility interdependence.[32] States have preferences not only about what they want others to do but about what others will obtain. Nations are not unlike people who have preferences not only about their own income but about the societal distribution of income. Not only do states want more power for themselves, but they care about the distribution of power. At any point they want to see their rivals weakened and their allies strengthened. Dilemmas of strategic choice also arise, therefore, when states must choose between their selfish interests and their extended preferences regarding the payoffs of others. In competitive bilateral relations this dilemma is between maximizing absolute and relative payoffs; in allied relations it is between absolute and conjoint payoffs. These represent dilemmas rooted in alternative decision criteria.

Such competing logics of choice—the existence of different bases for calculating interests—entail alternative conceptualizations of opportunity costs. Just as actual gains can be assessed in different ways, so too can forgone gains. When, for example, a competitive logic conflicts with an individualistic one, either choice will entail either absolute or relative opportunity costs.

One implication of this book is that more can be said formally and deductively if some of the usual narrow assumptions, drawn largely from economics, are relaxed. Actors do not merely transform their own payoffs into utilities and maximize them. Assumptions of selfish rationality, maximization, and full information are too confining. Actors misperceive. They employ competitive and conjoint decision criteria as well as solely self-regarding ones. They differ in attitudes toward risk as a function of their status quo position and their expectations.

One certain implication of relaxing assumptions in further deductive work will be the proliferation of dilemmas. These results mirror the literature that scholars have generated in a search for mechanisms to aggregate individual preferences into social choices. That field is littered with impossibility theorems and an array of solution concepts. On the one hand, it is easy to generate a set of criteria and demonstrate that there exists no solution that satisfies all the criteria (i.e., impossibility theorems). Alternatively, different criteria can be relaxed to generate a set of outcomes that meet the remaining criteria.[33] The same is

---

32. For a compilation of works from various disciplines that include such conceptualizations, see Charles G. McClintock, "Evolution, Systems of Interdependence, and Social Values," *Behavioral Science* 33 (January 1988): 59–76. His list (especially as categorized in Figure 1) should be supplemented with various items cited in Chapters 5 and 6.

33. In addition to any introduction to social choice, see Martin Shubik, *Game Theory in the Social Sciences: Concepts and Solutions* (Cambridge, Mass.: MIT Press, 1982).

clearly true of strategic interaction. One can generate multiple criteria and demonstrate that situations exist in which not all can be satisfied. The criterion of dominance, for example, clashes with that of optimality in the prisoners' dilemma.[34] Alternatively, one can play with the assumptions and generate new equilibria.[35]

Dilemmas Rooted in Asymmetries

Two asymmetries bear on dilemmas of strategic choice. First, the extant power and resources that states bring to any situation vary. These differences become enshrined in the distinctions among hegemonic powers, great or major powers, and others. Second, the payoffs from any interaction can themselves be asymmetric. Although two states may both most prefer the same outcome, their actual payoffs from it can be quite different.

In part, the variable nature of payoffs is a function of the differences between states themselves. Powerful states are more capable of absorbing losses than are weaker ones. Hence the former can afford to emphasize the long term, and they can take risks that more fragile nations dare not.

When payoffs are asymmetric, whether owing to the characteristics of nations or in and of themselves, states may be forced to choose on the basis of either their own returns or the impact their choices will have on others. For competitors, this means choosing between absolute and relative standing; for allies, between the needs of the other state(s) and one's own absolute payoffs.

The analysis in the foregoing chapters shows that some dilemmas arise from the asymmetrical impact of one nation's decision. Dilemmas of rivalry and alliance can both occur when a state's choice has a greater impact on another country's outcomes than on its own. Only then does a nation confront a choice between maximizing its absolute and relative

---

34. A great deal of philosophic interest has been sparked by Newcomb's problem and the conflict between dominance and expected utility in that situation. For an excellent recent compilation that is a good starting point, see Richmond Campbell and Lanning Sowden, eds., *Paradoxes of Rationality and Cooperation: Prisoner's Dilemma and Newcomb's Problem* (Vancouver: University of British Columbia Press, 1985).

35. New constraints have been added to reduce the number of equilibria in extensive-form games. For a recent review, see Roger B. Myerson, "An Introduction to Game Theory," in *Studies in Mathematical Economics*, ed. Stanley Reiter (Washington, D.C.: Mathematical Association of America, 1986), pp. 1–61. For a new notion of equilibrium in two-person games, see Steven J. Brams and Donald Wittman, "Nonmyopic Equilibria in 2 × 2 Games," *Conflict Management and Peace Science* 6 (Fall 1981): 39–62.

payoffs. The same is true for an alliance member confronting a choice
between maximizing absolute and joint payoffs. The ability to have a
greater impact on another nation's payoff than on one's own can be
the basis for dilemmas between individualistic and nonindividualistic
decision criteria.[36]

Dilemmas rooted in asymmetric consequences are especially problematic for great powers, which can afford better than lesser ones to
abjure absolute short-term payoffs in favor of long-term and either relative or conjoint payoffs. Just as the stakes of any gamble matter less
to the rich than to the poor, so the consequences of any strategic choice
will likely have a greater impact on lesser powers than on great ones.
The luxury of strength, therefore, presents the greater power with the
dilemma of needing to decide whether to emphasize its own returns
or the impact of its choice on others' payoffs. The weaker state, however, less able to forgo gains in order to affect others' payoffs, will more
likely consider only its own returns. Although this predicament can
technically exist for weak nations, only for great powers does the choice
between relative and absolute power present a compelling problem, a
hegemon's dilemma.

Competition and Positional Proximity

Decisions can be made solely by reference to one's own payoffs or can
include the payoffs to others in the calculus. Solely self-interested actions can be taken exclusively on the basis of immediate returns or with
an eye to long-term payoffs as well. When others' payoffs matter, they
can matter positively or negatively. Rivals and competitors can be interested in maximizing the difference among themselves, whereas allies
may be interested in maximizing the joint returns to themselves. Moreover, the decision to cooperate or conflict depends not only on options
and payoffs but also on the ways in which the payoffs are assessed.
Whether one approaches an interaction as an individualist, competitor,
or ally affects one's choice—that is, the underlying relationship within
which strategic choice is made is critical.

36. Relational power—the ability to affect another's payoffs—can be distinguished
from metapower—one state's ability to structure another's options and payoffs. See T.
Baumgartner, W. Buckley, and T. Burns, "Relational Control: The Human Structuring of
Cooperation and Conflict," *Journal of Conflict Resolution* 19 (September 1975): 417–40,
and Tom Baumgartner, Walter Buckley, Tom R. Burns, and Peter Schuster, "Meta-Power
and the Structuring of Social Hierarchies," in *Power and Control: Social Structures and
Their Transformation*, ed. Tom R. Burns and Walter Buckley (Beverly Hills, Calif.: Sage
Publications, 1976), pp. 215–88.

This raises the question of what conditions underlie the use of competitive or conjoint decision criteria. As discussed in Chapter 5, certain issues (such as territory) naturally generate competitive assessments. But in other cases the development of a rivalry focused on relative gains is not inherent in the issue. The distinction between arms in the hands of one's allies and one's enemies demonstrates the different ways the same items can be viewed depending on the underlying relationship.

Moreover, typically the same decision criterion cannot simultaneously be applied to more than one state. This means that there is a metadecision preceding the actual decision. In effect, actors choose the criteria they apply in a particular relationship and then assess payoffs and make concrete decisions given that earlier choice. The metadecision—whether to relate to another as ally, competitor, or individualist—can be critical to the prospects for cooperation and conflict.

My own argument is that competition arises between states that are *positionally proximate*. Competition is not global, since states cannot compete with all others simultaneously and not all others are relevant or salient.[37] States are *proximal competitors*, they compete locally, with others who are similarly situated, whether geographically, geopolitically, militarily, economically, or in some other way.[38]

States are concerned with their most immediate neighbors. Before the modern era of airplanes and rockets, nations could be threatened either by states with which they shared borders or by those occupying and controlling these immediate neighbors. In most eras, therefore, and in many parts of the globe today, competitive orientations involve those nearby. Israel makes relative military assessments not vis-à-vis the United States or the European great powers, but with other Middle Eastern countries. Precisely this logic lies at the root of the argument that nuclear proliferation would occur in regional pairs—if India ob-

37. Paul Anderson and Timothy J. McKeown argue that states initiate war only against others with whom they have a history of substantial interaction, a factor they label "salience"; see their "Changing Aspirations, Limited Attention, and War," *World Politics* 40 (October 1987): 1–29. For a technical demonstration that the impact of relative gains decreases with an increase in the number of states in the system, see Duncan Snidal, "Relative Gains Don't Prevent International Cooperation," paper presented at the Workshop on International Strategy, University of California, Los Angeles, February 8, 1990. Yet as Leif Johansen shows, competitive concerns remain a problem between coalitions. A concern about the division of a surplus can mean that there will be no core in cooperative games. See Johansen, "Cores, Aggressiveness, and the Breakdown of Cooperation in Economic Games," *Journal of Economic Behavior and Organization* 3 (1982): 1–37.

38. My formulation of competition as rooted in positional proximity should be compared with the factors listed by Stephen M. Walt as determinants of threat; see Walt, *The Origins of Alliances* (Ithaca, N.Y.: Cornell University Press, 1987).

tained the bomb, so would Pakistan, and so on. Proliferation in the third world would involve competitive regional relationships.

But in a day and age of planes and rockets, one state can threaten another from far away. The evolution of weapons systems has entailed increasing the distance over which lethal force can be applied. The range of guns exceeds that of knives, cannons that of guns. Ships, planes, and rockets can deliver firepower at long distances from their home bases.[39] Although competition in such a world could be global, states vie only with those of comparable strength and power. The United States, a major nuclear power, competes militarily not with Argentina, for example, but with the Soviet Union, the only state with the ability to threaten its survival. Although Argentina can be destroyed by both the United States and the Soviet Union, it is so much weaker that it cannot seriously compete militarily with either. Competitive military assessments involve states that are militarily comparable, states near one another on a global hierarchy of military power.

The same argument, that positional proximity drives competition, operates in the realm of economics. The United States, the world's premier economic power following World War II, has continually focused on the economic challenge of the most proximate economic power. In the 1950s it feared the growth of Soviet GNP. It worried little about the Japanese economy until it grew to become the world's second largest. Since then, the United States has increasingly focused on Japan as an economic competitor.[40]

The competition of positionally proximate states explains the nature of alliance dynamics as well. Alliances between positionally distant states, as between great powers and lesser ones, are nonproblematic. The United States, as discussed in Chapter 6, could even transfer an intercontinental weapons delivery system to another nuclear power, Great Britain, whose much weaker position meant that the United States did not even consider it a potential rival. But alliances among approximately equal great powers in multipolar worlds are much more problematic. Such alliances combine powers who could be rivals except

39. See the discussion of viability and the loss of strength gradient in Kenneth E. Boulding, *Conflict and Defense: A General Theory* (New York: Harper and Row, 1962).

40. Similarly, individuals compete economically with those nearest them on the economic ladder. Individuals do not typically feel disadvantaged by the successes of those far richer or far poorer than themselves, but focus on the relative economic success they enjoy in comparison to those of similar station. An associate professor unperturbed by the raise obtained by an extremely well-paid full professor will take great exception if someone of the same rank obtains a greater pay increase than the associate professor does. Competition is driven by positional proximity.

for the existence of a transcendent common rival. In such cases, although states may ally with other great powers, they will likely continue making individualist rather than conjoint assessments of interest.[41]

Force of Circumstance

In this book I have emphasized the importance of context. That element of context most likely to change from interaction to interaction is the payoff environment, the combination of potential returns. At times, a knowledge of just the actors' preference orderings is sufficient to predict choice, the nature of interaction, and the subsequent outcome. Sometimes, however, the actual size of the payoffs, not just their ranks, may be determinative. In the iterated prisoner's dilemma, for example, the actual magnitudes of the returns and the weights the actors attach to future payoffs matter enormously.

Strategic choice entails more than just a configuration of payoffs, however. Every interaction takes place against some background. States may be more or less powerful than one another. The outcomes of their previous interactions may also have consequences.

Choices are also made to change a status quo and to alter an expected trajectory. The status quo may be acceptable and desirable to some states, but not to others, which want to change it. Some states perceive time to be on their side, but others see their situations as deteriorating. These assessments have an impact on states' dilemmas and on the criteria of choice that states adopt to resolve them. They especially determine states' attitudes toward risk.

Cooperation and conflict are rooted in an interacting set of situational factors. International interactions occur in a context of some relationship of rivalry, conjunction, or neutrality. They occur against some backdrop of past history. They entail an international structure and a set of payoffs. In some cases the imperatives of structure, situation, and relationships can determine whether cooperation or conflict results.[42]

41. For recent work on states' concerns with the externalities of relationships, see Joanne Gowa, "Bipolarity, Multipolarity, and Free Trade," *American Political Science Review* 83 (December 1989): 1245–56, and Lars Skalnes, "Allies and Rivals: Politics, Markets, and Grand Strategy," Ph.D. dissertation, University of California, Los Angeles, in progress.

42. Jack Hirshleifer continually emphasizes issues of context. For his view of the importance of what he calls protocols of play, see Hirshleifer, "Protocol, Payoff, and Equilibrium: Game Theory and Social Modeling," University of California, Los Angeles, Department of Economics Working Paper #366, March 1985. In another paper Hirshleifer

Different Paths to Cooperation and Conflict

A myriad of contextual and situational factors can affect the very natures of the international conflict and cooperation that they generate. One implication is that an accurate understanding of international politics requires greater disaggregation and a more refined categorization of institutions and behaviors. Neither all cooperation nor all conflict is of a piece; there exists a diversity of each. Different cooperative and conflictual strategies are appropriate in different situations and circumstances. Regimes based on coordination and on collaboration are not the same. For one thing, they involve varying degrees of institutionalization. Not only do the forms of cooperation and conflict differ; so, too, do the routes to each. Alternative paths can lead to a single outcome.[43] Cooperation and conflict can emerge from the underlying preferences of states, but they can also result from nations' misperceptions of one another. They can emerge from individualistic assessments of payoffs or from the use of conjoint or competitive decision criteria.

Pure conflict, for example, can evolve in different ways. Zero-sum games can result from competition over a fixed good, such as territory, or because the nature of the good, such as a rival's military arsenal, itself generates the use of competitive decision criteria.

Cooperation, too, has different bases. In some cases it emerges from individualistic assessments of payoffs. At other times cooperation evolves as a response to the existence of competition in the international arena and comes to encompass bilateral relations of true helping behavior in which states maximize joint interests and attach weight to their allies' concerns. Such concerted choice represents more than temporary marriages of convenience and unbridled individualism.

A variety of contextual and situational factors matter and must be incorporated in empirical analyses of world politics. Situations with comparable payoffs may result in different outcomes because of divergent perceptions and decision criteria. Conversely, different constellations of payoffs can lead to similar outcomes for the same reasons. An

---

argues that conflict is a function of the interaction of preferences, opportunities, and perceptions. Opportunities can range from complementary to antithetical and preferences from benevolent to malevolent. See Hirshleifer, "The Expanding Domain of Economics," *American Economic Review* 75 (December 1985): 63.

43. Foreign policies often lumped together are arrived at in different ways. For interesting game-theoretic demonstrations of this, see George Quester, "Six Causes of War," *Jerusalem Journal of International Relations* 6 (1982): 1–23, and George W. Downs, David M. Rocke, and Randolph M. Siverson, "Arms Races and Cooperation," *World Politics* 38 (October 1985): 118–46.

awareness that particular outcomes are consistent with a multiplicity of causal paths epitomizes what deductive logic can bring to inductive work.[44] To be complete, empirical studies must deal with all the factors that determine strategic choice.

Theoretical and Policy Implications

The fact that there are different routes to conflict implies that there are different possibilities for conflict resolution. Arms races, which exemplify the creation of pure conflict over a variable good, provide an ideal illustration. Competitive arms escalations reflect rivalry, but not over a fixed good. As a result, one way in which the competition can be superseded is by the emergence of a transcendent threat. A country that feels itself threatened by another's procurements can find that the same foreign arsenals supplement its security when another, still more powerful and threatening rival emerges. In contrast, conflicts of interest rooted in the payoff environment itself, including rivalries over fixed goods such as territory, are not resolved merely by the emergence of another threat. They remain conflicts in need of resolution, although that process may be simplified if a common threat emerges.

Because different routes are taken to a particular outcome, it is difficult for one nation to unravel the bases of others' behavior—their intentions and their motivations. A state must determine whether others' actions reflect their dominant strategies or their contingent calculations, individualistic self-interest or relativistic competition, misperception or not. In the prisoners' dilemma, for example, defection can reflect greed or fear of exploitation by others.[45]

---

44. Interaction occurs within a context that includes irreversible choices. A framework of repeated interaction can be conceptualized as entailing reversible choices or not. Most economic analyses assume that actions are reversible and that consequences do not depend on the sequence of preceding events. Others see the world as path-dependent: actions are not reversible, and outcomes critically depend on choices made, timing, and the sequence of events. International relations obviously involve both phenomena. States can defect from a trade regime by establishing duties in the hope of renegotiating an agreement, in which case they will subsequently lift that duty. An ambassador can be recalled and then sent back. Diplomatic relations can be opened, broken, and reopened. A dead hostage, on the other hand, cannot be returned to life. Nor can a nation undo its attack on another country by withdrawing its forces.

45. The state may be an inveterate defector or have a contingent strategy. Moreover, defection can be the product of individualistic or competitive assessments. A state that conflicts with another cannot assess whether the other is acting on the basis of competitive assessments or individualistic ones. Similarly, a state cannot assess whether its allies are driven by individualistic self-interest or a concern with joint gains. Assessing intentions requires knowledge of others' payoffs, perceptions, calculations, and expectations.

The existence of multiple paths to conflict has made it possible for there to be arguments about how to prevent such discord.[46] Some scholars see the key in ending misperception. Some others believe it necessary to find alternatives to threats as a way of dealing with grievances. And still others, who view conflict as rooted in the structure of an anarchic system which generates conflicts of interest, find the solution lies in some form of overarching world government.

That there exist different routes to cooperation and conflict means that not only conflict but also conflict resolution is context-dependent. There can be no magic panaceas for resolving discord. What will work in some situations will not do so in others. The development of mutual understanding may result in cooperation or the absence of conflict, but not necessarily. This is because not all conflicts of interest are rooted in misunderstanding, and because reducing misperceptions can exacerbate conflict as well as alleviate it. Some conflicts can be resolved by complete information. Others cannot. Familiarity, after all, can breed contempt.

In short, there are different bases for conflict resolution, ranging from changing perceptions to altering the payoffs to changing the decision criteria. States can prevent conflict and ensure cooperation by emphasizing those logics that lead to the cooperative choice and deemphasizing those logics that lead to the adoption of a conflictual strategy.

Decision Criteria and Debates

Policy debates also reflect the existence of dilemmas. Foreign policy disagreements rarely occur when the international environment generates unambiguous imperatives and interests.[47] In such cases the adoption of any decision criterion, any temporal perspective, can lead to the selection of only one strategy.[48] Policy debates typically entail

46. To go with his six causes of war, George Quester provides six steps toward peace; see Quester, "Six Causes of War."

47. This point is made by Jervis, *Perception and Misperception in International Politics*, pp. 19–20. He goes on to argue that perceptions therefore matter. Although I do not disagree with this, my point here is that domestic policy debates can occur because of competing decision criteria even when all agree in their perceptions of payoffs and options.

48. One empirical study of the effects of personality on foreign policy position adopts the criteria of selecting only situations in which there were disagreements over policy at the highest levels of government. This controls for systemic effects because officials confronting the same international environment, presumably with access to the same information, are arriving at different conclusions as to what to do. It is in such situations, one scholar argues, that personality determines policy choice. Etheridge, "Personality Effects on American Foreign Policy, 1898–1968." Situations that entail dilemmas of strategic choice do, indeed, exist.

compelling logics for alternative strategies, however.[49] It is possible to agree on the payoffs associated with different outcomes and yet dispute the choice of a strategy. The disputants can all have the national interest at heart and base their cases on calculated and purposive judgments about it. But the use of different decision criteria and different temporal horizons can lead them to make different policy recommendations.

As the debate between liberals and mercantilists so aptly illustrates, not all policy disputes are resolvable merely by obtaining a consensus on values, objectives, payoffs, and perceptions. Advocates of promoting global welfare, proponents of maximizing national wealth, and champions of increasing relative wealth may all see their positions as upholding the same values and objectives in that all perceive themselves as proposing the best means of furthering the national interest in this domain. Still, their disagreement about the appropriate decision criterion leads them to make these divergent policy recommendations.

The arguments developed here about decision criteria and the nature of self-interest help to make sense of foreign policy debates. There has been a long-standing argument, for example, about U.S. foreign aid and whether it reflects America's national interests or whether it is driven by humanitarian concerns. Understood in the light of the arguments presented in this book, the debate is actually about how the United States calculates and assesses its interests. Indeed, the existence of the argument itself suggests that there is a dilemma, that use of an individualistic self-oriented criterion of self-interest leads to one strategy, whereas the use of a conjoint criterion that attaches weight to others' payoffs leads to the adoption of a different strategy. Those who argue for a self-oriented assessment of interests deride others as humanitarians. Those who emphasize conjoint interests either stress a broader notion of national interests or attempt to find an egoistic and instrumental basis for assisting others.

Domestic disputes between arms controllers and arms racers also reflect the divergent decision criteria used by the different sides. Arms controllers, of course, maintain that arms control is in the national interest. Typically, they argue that the prisoners' dilemma of arms procurements is Pareto-deficient and that arms control represents a Pareto-superior outcome that its mutually desirable. Arms racers point

49. For an argument that the policies advocated and chosen when there is an internal debate reflect both cognitive and organizational processes, see Lori Helene Gronich, "Expertise, Naiveté, and Decision Making: A Cognitive Processing Model of Foreign Policy Choice," Ph.D. dissertation, University of California, Los Angeles, in progress.

to the inherent problems in verifying an agreement and the difficulty of trusting the other side when it has an incentive to cheat. But they also emphasize the importance of relative power, the utility derived by the other side from an agreement, and the prospects that an agreement desired by the other side is probably unbalanced to the detriment of the United States.[50]

I have argued above that the debate between proponents and opponents of war-fighting strategies of escalation dominance also reflects a difference in how they assess decision criteria in nuclear crises. All agree that such situations are games of chicken, but they disagree about whether actors in such crises will focus on relative payoffs. If nations in a nuclear crisis are likely to focus on who has more to lose, it makes sense to develop forces for escalation dominance. If, on the other hand, nations will not make such comparative assessments, and if the differences between horrendous levels of casualties (the difference between fifty million and a hundred million casualties, for example) are not meaningful, escalation dominance is meaningless. The debate is essentially about the decision criteria that national leaders would use in a nuclear crisis. Both sides in the debate presume the rationality of decision makers in such a situation. They merely disagree on the bases of calculation that underpin rationality in these circumstances.[51]

In many cases debates about American foreign policy between right and left and between nationalists and internationalists reflect the use of different decision criteria to assess national preferences. When competitive and individualist criteria generate different strategies, the right emphasizes the importance of competitive considerations and the left emphasizes individual interests. Where individual and conjoint assessments diverge, nationalists stress the former and internationalists emphasize the latter.

50. For an empirical study that emphasizes the similarities of opposing arguments in national debates on armaments, see Fredrik Hoffmann, "Arms Debates—a 'Positional' Interpretation," *Journal of Peace Research* 7 (1970): 219–28. The article finds that the proponents of arms control make the classic arguments in its favor. Those advocating more weapons emphasize the importance of strong defense and the responsibility of other nations for the arms race. Missing in this analysis, and what I would venture, would be found in a reanalysis of these debates, are references to relative position.

51. James DeNardo makes the case that people's views of President Reagan's Strategic Defense Initiative (Star Wars) reflect their different assessments of the constellation of preferences in American-Soviet relations. He develops a theory of the roots of these as a function of different views of the nature of weapons and of deterrence. See DeNardo, "The Structure of Preferences in the SDI Debate," University of California, Los Angeles, working manuscript. I would only hold that such differences can arise either because of disagreements about the actual payoffs or because of disagreements about the decision criteria used to translate payoffs into preferences.

When a state incorporates the payoffs of other nations in its own calculations, those other countries are put in the position of defining its interests for it. When a state maximizes its relative position, what hurts its rival becomes important to it. Moreover, what is good for its rival is bad for it. In short, its own interests are defined in contrast with those of its rival. This has been a major criticism of American foreign policy in the years since World War II. American interests in the immediate postwar period were focused on Western Europe. But the Korean invasion put Asia on the Cold War map. Although American policymakers had not included Korea within the American defense perimeter in 1950, the United States intervened militarily in response to the invasion. Similar arguments, that the United States has allowed Soviet actions to determine American interests, for instance, have been made about Vietnam and Afghanistan as well.

A state that incorporates its ally's needs in its own calculations allows that ally, at least in part, to define its interests. This, too, is an argument that has been made about American foreign policy in the postwar period—that its allies have embroiled the United States in policies contrary to its own interests.[52] The implicit argument is that the conjoint calculation of interest diverges from the individualistic assessment of self-interest and that the relationship is not worth the cost.

When the use of different decision rules generates conflicting strategies, domestic debates regarding allies and rivals can revolve around the appropriate criteria to use in relations with them. As regards rivals, the debate is between those pressing for a policy that maximizes relative gains and those arguing for an individualistic policy that maximizes absolute gains. As regards allies, the debate is between those recommending a policy that reflects conjoint interests and the importance of the relationship and those emphasizing egoistic self-interest and absolute payoffs. Each side in such disputes sees itself as the repository of the national interest. What is at issue is the appropriate means for assessing self-interest.

That there are multiple criteria for calculating interests is evident in the mere existence of exhortations about the need to pursue the national interest. Modern realism is rooted in hortatory assaults on the failure of the Western powers to resist aggression in the 1930s. Exhortations to define the national interest in terms of power suggest that nations either may not know or may not pursue their interests.[53] It is

52. The most recent example of such an argument is that Israel ensnared the Reagan administration into selling arms to Iran.
53. Realists constantly argue on behalf of pursuing the national interest in domestic

also ironic that realists emphasize that systemic forces drive state policy even as they offer prescriptions for state policy. If structure generated an unambiguous imperative, scholarly injunctions would not be necessary.[54] Realists engage in domestic debates about how to assess and calculate interests, but the mere existence of such debates and of realists' scholarly injunctions to policymakers demonstrates the inadequacy of realist explanations.

When nations confront dilemmas of strategic choice, policy debates focus on the criteria by which decisions should be reached. Precisely because compelling logics exist for alternative courses of action, such disputes are not amenable to empirical or analytic resolution. There are neither empirical nor analytic means available for deciding, for example, whether a short-term or a long-term choice makes more sense in a particular case.

In explaining a nation's choice, domestic politics becomes critically important when ambiguity remains after structural factors have been considered. When structure provides no clear behavioral imperative, internal political debates arise, and their resolution determines the state strategies adopted. In other words, domestic political processes and the nature of political systems play a determinative role in such cases.

This book is built largely on the billiard-ball model of international politics, which treats states as unitary actors. In fact, it uses the words "actor" and "state" interchangeably. Yet nations are complex, themselves composed of interacting individuals and corporate entities. This does not matter if one set of these components defines interests and selects options. It may matter if a domestic game is embedded in an international one, if foreign policies entail not only interaction with other states but interactions within the state as well. There can be debates over how to assess national interests, about appropriate deci-

---

debates even as they argue that realism represents a positive theory of state behavior. Note the irony in the following titles: Hans J. Morgenthau, *In Defense of the National Interest* (New York: Knopf, 1951), and Stephen D. Krasner, *Defending the National Interest: Raw Materials Investments and U.S. Foreign Policy* (Princeton, N.J.: Princeton University Press, 1978).

54. As the title of his book (*Structural Conflict*) suggests, Stephen D. Krasner argues that conflict is built into the structure of international politics and particularly into the relations between the third world and the first world. Yet he exhorts people to recognize that the third world is pursuing power and that the rich nations misconstrue the true interests of the third world. See Krasner, *Structural Conflict: The Third World against Global Liberalism* (Berkeley and Los Angeles: University of California Press, 1985). But most important, implicit in the work is the assumption that there exist multiple criteria for assessing self-interest and that, therefore, conflict is not structural at all.

sion criteria, and about temporal horizons. There may be divergent perceptions of other nations' motivations and intentions. Policy may well be the result of a complex meshing of domestic and international relationships, therefore. Just as alliances cannot be understood apart from their being situated in a larger context of conflict, so, too, outcomes in international settings may not be fully explicable or comprehensible without attention to domestic interactions. Indeed, understanding the disaggregated nature of the nation-state has led countries to adopt foreign policies in order to strengthen the moderates or soft-liners and undercut the radicals or hard-liners in some foreign regime.

The fact that cooperation and conflict can result from the interaction of different contextual factors is also the basis for alternative interpretations of important historical events. Scholarly disagreements often revolve around the different routes by which a conflict can emerge—whether, for example, war is the product of misperception, of structurally based conflicts of interest, of an aggrieved actor's threats gone awry, or of competitive transformations of benign payoff environments. Each of these can be the basis for conflict, and each is consistent with its emergence.

Some scholars view World War I as a product of German aggression. Put differently, they implicitly argue that the Germans had a dominant strategy of defecting and so must have been willing to go to war if others chose to fight rather than capitulate. Others perceive German defection as having been contingent on other nations' decisions and as having resulted from a misperception of those others' preferences. This suggests that better signaling by the other European states might have prevented war. Even those who agree that Germany did want war dispute whether its defection was driven by rapaciousness or insecurity. If the former, the question arises whether Germany could have been dissuaded by toughness. If the latter, however, conciliation might have kept the peace.[55]

Resolving such historical disputes requires a good deal of information. Payoffs, perceptions, and decision criteria all interact to underlie a strategic choice. A particular choice can represent a dominant strategy but be based on more than one set of decision criteria or reflect contingency and misperception.

55. Robert Jervis makes a similar point in distinguishing the spiral model from the deterrence model as roots of conflict; see *Perception and Misperception in International Politics*, chap. 3.

Domestic Political and Economic
Considerations

The importance of domestic political and economic considerations is especially unappreciated in the areas of grand strategy and national security policy.[56] National security policy represents a nation's politico-military response to the dangers it confronts in an anarchic environment composed of other nation-states. Such policies necessarily respond to the challenges of the international system, but they also represent reactions to the constraints and pressures of domestic society. All too often, however, national security policy is analyzed as if it were fully unaffected by domestic constraints—as if it represented an optimal response to external military pressures. But unless structural constraints force a specific choice, systemic factors circumscribe a set of possibilities rather than determine a specific one.

The implications for national security policy of domestic politics, and especially questions of political economy, can easily be seen in discussions of arms races, which are typically bilateral competitions that occur as two states respond to each other's increased military procurements. Clearly, the actions of one state drive the responses of the other state. But just as obviously, the very existence of an arms race presumes that each state's domestic political economy makes increased defense spending possible. After all, the provision of national defense requires the existence of extractable domestic resources. The level of a nation's material wealth, the health of its economy, and its political ability to mobilize resources all play a key role, therefore, in determining the nature and degree of military effort. In an arms race each nation must either have growing state revenues or be able to increase its degree of revenue extraction, shift resources from other programs, or borrow additional funds.[57] Unless both nations have such an accommodating domestic political economy, an arms races will not occur.[58]

56. This section draws upon Arthur A. Stein, "The Political Economy of National Security Policy," paper presented at the annual meeting of the American Political Science Association, Atlanta, September 1, 1989, and "The Political Economy of Grand Strategy," paper presented at a conference with the same title, University of California, Los Angeles, March 17, 1990.
57. Arms races between minor powers can also be sustained by arms shipments from their major-power allies.
58. This puts much of the empirical work on arms races in a new light. Early studies that modeled one side's arms expenditures as a function of the other side's spending in the immediately preceding time period included a fatigue term in order to capture the obvious reality that arms races cannot continue indefinitely but are constrained by do-

Competitive armaments dynamics in the nineteenth century illustrate the importance of domestic political economy. Anglo-French naval races were short-lived owing to the domestic inability of the French to sustain them. France alternated between naval buildups and cutbacks because of its taxpayers' unwillingness to sustain expenditures.[59] In contrast, Germany entered a naval race against Britain, even in the face of fiscal constraints, because of domestic pressures. The Navy League, a large interest group mobilized on behalf of the navy, was founded with financial support from Krupp, a German steel and arms manufacturer, which had lost its export market to competitors.[60] In Great Britain, domestic politics enabled the government to sustain increased naval expenditures, especially during domestic economic slumps. Depressions had, until the 1880s, generated pressures for reduced government spending, including defense outlays. When the franchise was extended, Parliament came to represent not only property owners and taxpayers, but also those who either worked but did not pay taxes or were unemployed. This changed environment meant that depressions brought pressure for more defense spending. The poor did not pay the higher taxes, but they did reap the benefits. As a result, Parliament appropriated naval funding even during downturns; in some cases it even exceeded the military's spending requests.[61] The succession of political decisions that generated larger naval expenditures derived not only from international competition and rapid technological change but from the altered foundation of British domestic politics.[62]

The temporal horizon underlying state choices can also reflect the workings of the domestic political economy. Financial considerations, for example, constrained Britain's response to continental developments in the 1930s to one of short-term appeasement and long-term

---

mestic economic and political factors. Subsequent work modeled cost and resource constraints as restraining arms races. More recent studies have contrasted the relative explanatory power of domestic imperatives (usually measured by the previous year's defense expenditures) with reaction to the other side's spending. My argument is that accommodative domestic factors underlie the very existence of an arms race. A nation's participation in an arms race is inherently the product of the interaction of another state's expenditures with an accommodative domestic political economy.

59. William H. McNeill, *The Pursuit of Power: Technology, Armed Force, and Society since A.D. 1000* (Chicago: University of Chicago Press, 1982), pp. 227–28, 264, 299.

60. Ibid., pp. 301–4. The pivotal figure in focusing on domestic determinants of German foreign and military policy is Eckart Kehr, *Economic Interest, Militarism, and Foreign Policy: Essays on German History*, trans. Grete Heinz, ed. with an introduction by Gordon A. Craig (Berkeley: University of California Press, 1977).

61. McNeill, *The Pursuit of Power*, pp. 269–70, 275–76.

62. Ibid., p. 277.

deterrence. The British Treasury simply could not afford vastly expanded military expenditures. In addition, increased defense spending
would have entailed an increase in requisite imports and so would
have worsened the nation's balance of payments and its ability to maintain its international position. Together, therefore, domestic and international economic constraints precluded a military buildup, without
which Britain could not stand up to Hitler's advances and demands.
Appeasement reflected neither the British elite's admiration of Hitler
nor any British misunderstanding of the nature of his intentions, but
a financial inability to sustain any alternative national security policy.
By choosing a policy of short-term appeasement, Britain maintained as
strong a national economy as possible in order to provide a long-term
deterrent that would enable it to confront Germany with the prospect
of losing a war that included sustained mobilization.[63]

The position of the United States vis-à-vis the Japanese in 1941 was
somewhat comparable to that of Great Britain facing the German challenge in the 1930s. It did have, even in the eyes of Japanese leaders,
the ability to mobilize a vast military machine that would eventually
defeat Japan in a protracted war. In other words, the United States
possessed a *mobilization deterrent* against a Japanese attack. But it did
not possess an *immediate deterrent*, as evidenced by the Japanese decision to gamble that inflicting a major blow on the United States at
Pearl Harbor would convince the Americans not to wage the long war
that it could certainly win.[64]

In short, arms races depend on the existence of mutually accommodative domestic political economies, and successful deterrence in such contexts presumes symmetrical political economies and strategies. Arms
negotiations, if undertaken, will channel, but not end, arms races.

There may also be times when rivals simultaneously confront economic constraints, however, and in such cases both may moderate their

63. For an excellent discussion of the evolution of the literature on appeasement, see
J. L. Richardson, "New Perspectives on Appeasement: Some Implications for International
Relations," *World Politics* 40 (April 1988): 289–316, and see the items cited there in note
35 for the economic constraints on British foreign and defense policy. No one disputes
the importance of the Treasury in determining the course of defense budgets in the
interwar period; the disagreement is about whether there was an alternative. This argument about long- versus short-term deterrence is discussed in Chapter 4, and see Alan
Alexandroff and Richard Rosecrance, "Deterrence in 1939," *World Politics* 29 (April 1977):
404–24.
64. Bruce M. Russett, "Pearl Harbor: Deterrence Theory and Decision Theory,"
Journal of Peace Research 4 (1967): 89–105. See also Chihiro Hosoya, "Miscalculations
in Deterrent Policy: Japanese-U.S. Relations, 1938–1941," *Journal of Peace Research* 5
(1968): 97–115.

relativistic demands in such a fashion as to make arms control agreements viable.[65] Contemporary U.S.-Soviet relations, and the national security policies of each, may reflect the peculiar confluence of two societies confronting domestic political and economic constraints.

The world is changing and evolving. When territorial issues predominate, international conflict is the norm. But as more and more borders become mutually accepted, and as fewer states concentrate on territorial expansion, the prospects for more cooperative international relationships improve. The existence of nuclear deterrence in the modern world means that fewer states fear for their survival, and this, too, improves the prospects for long-term calculations that result in increased cooperation.

Moreover, the modern state is responsible for providing wealth as well as security. Especially in advanced industrial societies, citizens look to their governments to generate economic development and growth. They no longer see cycles of boom and bust as inevitable and outside the scope of state responsibility. This emphasis on ensuring wealth means an increasing concern with exchange that increases the likelihood of international cooperation.[66] Further, it means a lessened willingness to abjure absolute gains.[67]

Concluding Thoughts about the Power to Construct Social Reality

Cooperation is a product of choice and circumstance. Nations choose to cooperate when it is in their interest to do so, and it is the concatenation of forces of circumstance that shapes international affairs. In this book the existence of choice has been taken as given. Although structure plays a key role, however, it is not an unambiguous guide; it

65. In such cases policies of appeasement, conciliation, and negotiation will not lead to emboldening rivals and the failure of deterrence.

66. For an argument that economic exchange generates new sources of conflict which require greater cooperation to sustain further increases in exchange, see Richard Rosecrance and Arthur Stein, "Interdependence: Myth or Reality?" *World Politics* 26 (October 1973): 1–27, and especially Arthur A. Stein, "Governments, Economic Interdependence, and International Cooperation," in *Behavior, Society, and Nuclear War*, vol. 3, ed. Philip E. Tetlock, Jo L. Husbands, Robert Jervis, Paul C. Stern, and Charles Tilly (New York: Oxford University Press, for the National Research Council of the National Academy of Sciences, forthcoming).

67. This disinclination, particularly true in representative political systems, can interact with economic conditions. Thus it becomes more difficult, for example, to sustain a concern with joint gains during economic downturns (which generate an every-country-for-itself mentality). It is also difficult to sustain an emphasis on the long term during economic downturns.

is sometimes indeterminate and, therefore, incomplete. In such cases decision criteria are central, and here domestic politics loom especially large. So, too, can the creative discovery of new options and alternative conceptualizations define the very essence of policy and statesmanship. The world involves real constraints, but social choices and relationships are often malleable. Issues can be structured in various ways, and some impasses can be broken by creative formulations. It is not always possible to distinguish between what is given and what is perceived and created in the world.

The Role of Statesmanship

The concept of collective goods provides one example of the role of human ingenuity in the construction of social reality. Analysts can define such goods and then characterize particular items as collective or not. But the collective nature of a good does not always inhere in it. Sometimes it does; we cannot eliminate air pollution for some without also doing so for others. In other cases, however, collectivity is given by the nature of our institutions, policies, and conceptualizations. Governments can build freeways, or they can decide to collect tolls on limited access highways.

The history of international trade agreements illuminates the human creation (or social construction) of collective goods. Trade is inherently a private and noncollective good. The history of trade relations is replete with discriminatory arrangements between states. States can bilaterally negotiate tariff agreements that discriminate against other nations. They can differentiate between products. This makes it difficult to distinguish liberal and illiberal trade, for nations can, and do, pursue both simultaneously. A state can have unfettered trade with some partners and not with others. It can have no barriers on some products and maintain high ones on others.

Historically, trade agreements were mercantilistic instruments that discriminated against other nations. The free trade era and the ability to consider liberal trade to be a collective good emerged from the adoption of unconditional most-favored-nation clauses in bilateral trade agreements. Such clauses required the contracting states to extend to one another any concessions they later made to any third parties. When states bound by such agreements negotiated subsequent agreements that also included such clauses, they created a network of liberalizing trade agreements that provided a collective good in the sense that states within the set of signatories could not be excluded from the good.[68]

68. This discussion draws upon Stein, "The Hegemon's Dilemma."

The Role of Will and Belief

Social reality can also be affected by will and belief. The ability to see the world as it might be, as opposed to the world as it is, distinguishes the revolutionary from the pragmatist, the utopian from the realist.[69] The ability to transform the choices and payoffs of actors represents power—intellectual as well as military. Most of the time, wishing does not make it so. But delusional will can be efficacious and, therefore, rational. The dilemmas discussed in this book all emerge from purpose and calculation. Yet there can also be motivated (or calculated) irrationality.[70] When outcome is related to effort, the ability willfully to fool oneself can have an immense impact. Assume, for example, that two basketball teams are both twenty-five points behind at the half time.[71] They have been totally dominated by their rivals. At half time, one coach offers the truth: the team has been badly outplayed, and its odds of victory are infinitesimally small. The other coach excitedly declares that victory is at hand, for the team cannot possibly continue playing as badly, nor its rival as well, during the second half. The optimistic coach, explaining that the key is not to try to close the gap all at once, delineates the ways in which the team needs to improve its play. If the players do so, they will cut the lead down to twelve points halfway through the second half and to six points with five minutes to go; they will tie the other team at the two-minute mark, and they will win. The basketball coach who tells the truth is preparing the team for defeat. The coach who lies, on the other hand, motivates the team to play its best under difficult, if not impossible, circumstances. It is not probable that this team will win the game. But if once in every thousand times a team comes back from a deficit of twenty-five points, it is likely to be under the direction not of a coach who paints a realistic picture of the difficulties ahead, but of one who lies.

There are times when the will to win can make a marginal difference. The asymmetry of motivation can counterbalance asymmetries of power and determine the victor of a war.[72] Whenever a David defeats a Goliath, determination is likely to play a role. The power of positive thinking cannot overcome all impediments, but it is hardly inconse-

69. On pragmatic versus charismatic and revolutionary leaderships, see Henry Kissinger, "Domestic Structures and Foreign Policy," *Daedulus* 59 (Spring 1966): 503–29.

70. David Pears, *Motivated Irrationality* (Oxford: Oxford University Press, 1984).

71. I am indebted to my colleague Victor Wolfenstein for this example.

72. Outcomes of war are a function of a number of factors, and intangible ones can matter enormously. See Arthur A. Stein, *The Nation at War* (Baltimore: Johns Hopkins University Press, 1980), and Arthur A. Stein and Bruce M. Russett, "Evaluating War: Outcomes and Consequences," in *Handbook of Political Conflict: Theory and Research*, ed. Ted Robert Gurr (New York: Free Press, 1980), pp. 399–422.

quential if belief summons effort and effort is a determinant of outcome.[73]

Sometimes events derive from international and domestic pressures, and there is little that individuals can do to make a difference. But individuals can matter sometimes, even if only at the margins.[74] Martin Luther may be necessary to explain the Reformation, Gorbachev to explain recent events in the Soviet Union and Eastern Europe. But a Gorbachev makes a difference in a context—in a particular time and place and constellation of options and payoffs. His choices, too, are constrained by structure and circumstance, by the balance of terror and the domestic economy.

The story is told of the man who, hearing of an exchange of fire between Arabs and Jews in Jerusalem in the spring of 1921, donned his rabbinic robes and went to the field of battle.[75] He went out between the warring sides and asked that the firing cease. He spoke of their common love for the land and of brotherhood. When he was done, both sides dispersed and went home. The conflict did not end because of that one speech. Fighting resumed in the city in 1929. Underlying tensions were not put to rest. Indeed, as Charles Issawi describes the long-standing reasons that would-be peacemakers in the Middle East have for humility: "God sent Moses, and he couldn't fix it; he sent Jesus, and he couldn't fix it; he sent Muhammad, and he couldn't fix it."[76] Still, on that day in 1921, some men who otherwise would have died went home to enjoy life with their families. As we appreciate all too well from our experiences with modern medicine, the extension of life by even a few days is no small miracle, and one for which we are prepared to pay no small price. It is for this reason that people continue to analyze and argue and strive to make a difference.

73. See the discussion in Robert H. Frank, *Choosing the Right Pond: Human Behavior and the Quest for Status* (New York: Oxford University Press, 1985), pp. 30–32. Also see Shelley Taylor, *Positive Illusions: Creative Self-Deception and the Healthy Mind* (New York: Basic Books, 1989).

74. Philosophers have long debated the relationship between free will and determinism, or what Steven Lukes refers to as power and structure; see Lukes, "Power and Structure," in *Essays in Social Theory* (London: Macmillan, 1977). Nothing I say here should suggest that I accept what I would call the radical cognitivist position, that there is no reality except in the mind of the beholder. Deconstructionism in international relations, as elsewhere, strikes me as inherently marginal and constrained by structure and circumstance.

75. The story comes from Abraham Joshua Heschel, *Israel: An Echo of Eternity* (New York: Farrar, Straus and Giroux, 1969), pp. 175–78.

76. Charles Issawi, *Issawi's Laws of Social Motion* (New York: Hawthorn Books, 1973), p. 114.

Index

Library of Congress Cataloging-in-Publication Data

Stein, Arthur A.
  Why nations cooperate : circumstance and choice in international relations /
Arthur A. Stein.
       p.   cm.
  Includes bibliographical references.
  ISBN 0-8014-2417-8 (alk. paper)
  1. International relations.   2. International relations—Psychological
aspects.   3. International relations—Social aspects.   I. Title.
JX1391.S76   1990
327—dc20   90-55132